THE MYTHS OF INFORMATION

Theories of Contemporary Culture, Volume 2
Center for Twentieth Century Studies
General Editor, Thomas Ewens

THE MYTHS OF INFORMATION:
Technology and Postindustrial Culture

edited by
KATHLEEN WOODWARD

Coda Press, Inc.
Madison, Wisconsin

Library of Congress Cataloging in Publication Data
Main entry under title:

Myths of information.

    (Theories of contemporary culture ; 2)
    1.   Technology and civilization—Addresses, essays,
lectures. 2.   Information theory—Addresses, essays,
lectures.   3.   Mass media—Addresses, essays, lectures.
I.   Woodward, Kathleen M.   II.   Series.
CB478.M9   303.4'83   80-23653
ISBN 0-930956-12-5
ISBN 0-930956-13-3 (pbk.)

Printed in the United States of America

A co-publication of The Center for
Twentieth Century Studies, University
of Wisconsin, Milwaukee, Wisconsin &
Coda Press, Inc., 700 West Badger Road,
Madison, WI 53713
ISBN 0-930956-13-3

For
Michel Benamou

1929-1978

Preface: <u>The Devil's Hand</u>

> Once is happenstance
> Twice is coincidence
> Thrice is the Devil's hand

The present book evolved from two symposia on Technology and Postindustrial Culture sponsored jointly by The Center for Twentieth Century Studies at the University of Wisconsin-Milwaukee and the Centre du 20ème Siècle at the University of Nice. These two meetings—in Milwaukee in 1977 and in Nice in May 1978—were exemplary in many respects. It was a truly international venture, involving several hundred participants from many countries and highly diversified fields of research. Papers were on an extremely high level of scholarship and the meeting of so many original minds sometimes resulted in unforgettable exchanges. Major figures in the world of communications, sociology, culture and technology, literature and the arts, whose names appear on the following pages, encountered challenging response from unusually attentive audiences.

The outstanding nature of this research can be attributed to the fact that it was masterminded at Milwaukee's Center which Michel Benamou had in a few years made into one of the world's most fascinating crucibles of contemporary thought.

Michel and I met at the Sorbonne in the late forties. For the next twenty years we completely lost track of each other, until one day in 1973 when we ran into each other in Brasilia and suddenly realized that, without any contact during all that time, our lives had followed consistently identical patterns. The next year we each took charge of a research center, his in Milwaukee, mine in Nice, both bearing the same name, both founded in 1968 and both devoted to strangely similar objectives: listening to and identifying the voice of our declining century. Convinced that so many convergences could not be merely coincidental, we decided to team up our minds and our Centers and undertook an ambitious and far-reaching common research program whose first stage was to be the symposia on technology and Postindustrial Culture. Their unqualified success allows us to guess at what might have come out of further projects.

How could I have known, as we parted on the steps of the Palais des Festivals on a warm afternoon in May 1978, still excited by post-colloquial discussions, that Michel would die in Paris a few days later. Never have the words usually associated with death—"untimely," "outrageous," "absurd"—seemed so hopelessly inadequate to describe the blow borne us by the Devil's hand and the state of vacuum we were left in and from which I think none of us will every fully recover. Popular wisdom is mistaken: some men *are* indispensable and their demise leaves us pondering how much more meaningful history would have been had they survived.

A rare friend is gone. In lieu of an obituary, which Michel would have found ludicrous and irrelevant, I express my hope that this book will help acquaint scores of unknown friends throughout the world with Michel Benamou's vision of the earth we live on. Such is, I am sure, the wish of the many distinguished contributors who shared with us the adventure of these encounters and who feel that these pages, not the waters of the Marseilles harbor where his ashes are scattered, will keep the memory of an extraordinary man.

Michel Sanouillet
Director, Centre du 20ème Siècle
University of Nice, France

Acknowledgements

The origins of this book lie in the spirited conversations about technology and contemporary Western culture between the late Michel Benamou, Director of the Center for Twentieth Century Studies (1974–1978) at the University of Wisconsin-Milwaukee, and Michel Sanouillet, Director of the Centre du 20ème Siècle at the University of Nice. In 1977–78 they twinned the two research Centers and mounted a moveable feast of intellectual events, which Michel Sanouillet has described in the Preface. This volume—the second in the Center for Twentieth Century Studies series of *Theories of Contemporary Culture*—is a result of that heady year of collaboration. In Nice the Centre's new journal *Medianalyses* is also publishing many of the papers presented during that year. One in English, the other in French, these volumes are co-publications in the real sense of the word. They testify to our excitement in pursuing collective research in the humanities across the ocean and the even more obdurate barriers all too often erected in educational institutions.

At the University of Wisconsin-Milwaukee, we have indeed been fortunate. I admire the commitment of the College of Letters and Science and the Graduate School to research in the humanities and to this project in particular. The support of Dean William F. Halloran and Dean G. Micheal Riley has been especially generous. The assistance of the Graduate School, under the leadership of Dean George

Keulks, permitted us to invite Marie-Pierre de Montgomery, the Associate Director of the Centre in Nice, here in the fall semester of 1978. Her welcome residency enabled us to continue work on these cross-Atlantic publications.

Research in the domain of technology and culture is one of the major areas of inquiry of the Center for Twentieth Century Studies, at present under the direction of Thomas Ewens. Two other books are in the works: *The Cinematic Apparatus,* edited by Teresa de Lauretis and Stephen Heath (forthcoming from Macmillan) and *The Technological Imagination,* edited by Teresa de Lauretis, Andreas Huyssen, and myself, as the third volume in *Theories of Contemporary Culture.*

I am indebted to Teresa de Lauretis and Mark Krupnick, Acting Directors of the Center in summer 1978 and the academic year 1978–79 respectively, for their acumen and encouragement. The Center staff has consistently displayed the high level of expertise and friendship for which it has been known. I am grateful especially to Charles Caramello, Jean Lile, Mary Lydon, Paul Ogren, Josie Simmons, and Carol Tennessen.

I take pleasure in acknowledging the National Endowment for the Humanities and the French Cultural Services of New York and Chicago for their support of the International Symposium on Post-industrial Culture: Technology and the Public Sphere, held November 17–19, 1977, in Milwaukee.

I would like to express my personal appreciation to Carl Lovitt of Coda Press for his assiduousness in the preparation and publication of this book, to Roy Behrens for his cover design, and to Marc Benamou for his help with translation. Finally, I would like to thank Michel Sanouillet and Marie-Pierre de Montgomery and all the contributors to this volume for their patience, understanding, and rare affection in helping me complete what was not only an intellectual adventure but an affair of the heart.

Kathleen Woodward

TABLE OF CONTENTS

IV. CYBERNETICS, CONSTRAINTS, AND SELF-CONTROL

Introduction

Unfortunately, "information retrieving," however swift, is no substitute for discovering by direct personal inspection knowledge whose very existence one had possibly never been aware of, and following it at one's own pace through the further ramification of relevant literature. But even if books are not abandoned, but continue their present rate of production, the multiplication of microfilms actually magnifies the central problem—that of coping with quantity—and postpones the real solution, which must be conceived on quite other than purely mechanical lines: namely, by a reassertion of human selectivity and moral self-discipline, leading to continent productivity. Without such self-imposed restraints the overproduction of books will bring about a state of intellectual enervation and depletion hardly to be distinguished from massive ignorance.

Lewis Mumford,
The Pentagon of Power (1970)

The past thirty years have been characterized by intense speculation about the changing shape of industrial society. Book upon book of cultural theory and criticism, representing every conceivable discipline and inventing others, has responded to the passing decades: the emergence of the consumption society in the fifties, the heady utopianism of the sixties, and the sobering rediscovery of scarcity in the seventies. The level of thought has been highly creative; the vision, one of turning points; the tone, often apocalyptic. We have seen the proliferation of computers and the consolidation of the mass media. We have witnessed theoretical innovation in cybernetics, futures forecasting, and systems design. I need mention only a few writers to suggest the fertility and range of conjecture: Norbert Wiener, Lewis Mumford, Daniel Bell, Victor Ferkiss, Herbert Marcuse, Robert Heilbroner, and Buckminster Fuller in the United States; Marshall McLuhan and Harold Adams Innis in Canada; Jacques Ellul and Jean-Pierre Domenach in France; Raymond Williams and E. F. Schumacher in England; Jürgen Habermas in Germany.

The scope of such speculation has been extensive and its substance varied, but it is united in part by a concern with technology (or *technique,* as Jacques Ellul calls it) as an agent of cultural change. Indeed we now commonly refer to contemporary society as The Technological Society. By the very act of naming our historical moment "technological," we have identified technology as *the* motive force of change. It is as if we have only recently discovered technology, or technique, which was before largely invisible to us. This is an important insight, which has yielded much exciting research. We have studied technology in other cultures as well as in our own cultural past. We are now quite well aware of the fact that technology has always been an important dimension of human society. We have grown accustomed to think of man generically as a tool-maker. But our period seems qualitatively different. In order to understand our point in time, we have reified technology and invented new myths of history, of which we must be critical. Hence the title of this book— *The Myths of Information: Technology and Postindustrial Culture.*

A myth is a construct we invent to explain the unknown world to ourselves and, once upon remembered time, ourselves to the unknown world. Now, as then, myth is a reading of history in which is implicit the shape of things to come. Two such myths of technology, which are related to one another, constitute a subtext to this book. The first is the myth of postindustrial society itself, or the informational society, or the cybernetic society, or the technological society, as it is alternately called. For in order to distinguish *post*industrial society from industrial society, we have re-interpreted Western history as a sequence of three technological revolutions: the agricultural revolution, which is identified with tool-making and oral culture; industrial society, identified with the power technology of mechanization and print culture; and postindustrial society, with the informational technology of electronics, and what McLuhan has called the "literate orality" of mass media culture.

In his influential *The Human Use of Human Beings: Cybernetics and Society* (1950), Norbert Wiener announced the revolution which was being brought about by the development of computers and other communication and control systems. Daniel Bell, another influential spokesman for postindustrial society, has dated the point of visible change in 1956, the year in which the number of white-collar workers surpassed the number of blue-collar workers. To his mind, postindustrial society is marked by the liberation of the worker from the assembly line, the shift from manufacturing to services, and the new importance of theoretical knowledge to the production process. We see this most clearly in the new interdependence of science and technology, Bell has written, describing knowledge as the "ganglion" of postindustrial society.

The ganglion is the focal point of energy, the nervous source of strength. Others have shifted the metaphor to residually religious dimensions. Richard Hamming, for example, has extolled information as the "soul" of modern life. But if for Bell and others, knowledge and information are central to the production process, what do they produce? In one sense, it might be said that *they reproduce themselves.* Information is the distinctive commodity of postindustrial society. (We should remember that in 1949 Claude Shannon first proposed in *A Mathematical Theory of Communication* that information is a quantity which can be measured in physical terms. In the economic sphere, we now compute the information produced and add the total in to the gross national product.) Underlying the process of the reproduction of information are metaphors drawn from biology and fission, both of which proceed at an exponential rate. (I except here zero population growth, which is ideologically opposed to the myth of growth as progress.) We speak of an information explosion that triggers an ever-accelerating growth in information. The process is figured as being continuously fed by positive feedback, and the production curve of information charts a constantly increasing abundance. To many, this is a utopian vision, but to others, the information explosion means violence, destruction by information fallout.

The most well known of the prophets of postindustrial society is of course Marshall McLuhan, whose image of a global village united by communications networks and information-processing systems was

absorbed into Western consciousness in the sixties. McLuhan exul-
tantly proclaimed that the "'electric age' is recovering the unity of
plastic and iconic space, and is putting Humpty-Dumpty back together again." His prediction of the electric age was echoed by others,
in other fields. Here I mention only one: in *The Information Machines*
(1971), Ben Bagdikian writes of the social change being brought about
by the mass media and information-processing machines as "more
powerful than our innocent introduction to electronic pictures in 1927,
perhaps more significant than all past changes in the technology of
information." Bagdikian's tone is eager, revealing his own myth of
the informational society: before we were infants, now we are moving
toward maturity, a technological coming of age, our cultural prime.

No one would disagree that the globe is in fact "shrinking" as a
result of innovation in mass communication and transportation. Most
would agree that McLuhan's ecstatic vision of a "global embrace" is
very much a timebound expression of the sixties. The important issue
is just whose interests such a reading of history and the future serves.
For every myth serves an organizational and political function. Every
myth seeks to control and shape some aspect of a social system. Does
the myth of the informational society speak for a certain form of social
organization? Does it rationalize particular social practices? What, in
other words, is the ideology of the informational society? The very
term "postindustrial" seems to imply that the dominant form of
economic organization in the West has changed fundamentally. But is
this the case? One of the myths of the informational society is that in
the West we have moved beyond capitalism and its concentrations of
power, the exploitation of the worker and an emphasis on material
goods. As Jean-Pierre Dupuy points out in his essay "Myths of the
Informational Society," which is an excellent introduction to many of
the main themes of this book, the relationship between the production of goods and services may have indeed shifted dramatically in
the United States and other Western countries, but this is in part
because the nineteenth-century practice of colonialism persists. We
ship off much of our factory work to Third World countries. The
global village is a fiction of the relatively wealthy. Although capitalism
may now be more vulnerable to containment by the Third World, this
is primarily the result of two things: shifts in power due to the energy
crisis and the increased complexity and interdependence of the system itself which expose it to guerilla warfare.

If the upper limit of the size of a society is its capacity to store
information (assuming a society is not threatened from without), and
if we have significantly increased that ceiling through computer memory banks, microfiches, and other information technologies, we may

not have transformed the shape of Western society but rather pro-
longed its ability to perpetuate itself. This is what Joseph Weizenbaum
means when he observes in *Technology at the Turning Point* (1977) that
the computer came along just in time, allowing us to avoid revolutionary change by shoring up existing institutions and practices which
might have collapsed under the sheer weight of increasing data. A
better term to describe the character of contemporary Western culture
would be, in fact, the term Wiener introduced thirty years ago—the
Second Industrial Revolution—which acknowledges a fundamental
continuity in basic social and economic practice at the same time as it
recognizes significant technological change.

One of the most fascinating results of the invention of the computer and other communication systems is the permeation of other
realms of discourse by the vocabulary of information theory and data
processing, altering our vision of ourselves and the world. This brings
us to the second myth of technological change. As a touchstone, we
may again refer to the theory of McLuhan. If the first myth of the
informational society concerns the relationship of technology to the
nature of production, the second deals with the relationship of technology to the human body. Technologies are commonly regarded as a
means to extend, improve, and aid the body. A shovel is an extension
of the arm; a bicycle, an extension of the legs; the telephone, the
extension of the ear; the computer, the extension of the brain. Implicit
in this view of the development of technology is the belief that it will
reach its full expression in the informational society. Earlier I referred
to the information explosion. McLuhan's preferred metaphor, however, is implosion. As we read in *Understanding Media* (1964):

> After three thousand years of explosion, by means of fragmentary
> and mechanical technologies, the Western world is imploding. During
> the mechanical ages we had extended our bodies in space. Today, after
> more than a century of electric technology, we have extended our central
> nervous systems itself in a global embrace, abolishing both space and
> time as far as our planet is concerned. Rapidly, we approach the final
> phase of the extensions of man—the technological simulation of con-
> sciousness, when the creative process of knowing will be collectively and
> corporately extended to the whole of human society, much as we have
> already extended our senses and our nerves by the various media.

Having mapped the entire body with our technologies, we encounter
a conceptual limit. We cannot imagine a further advance, and the
appearance of electronic technologies seems a final stage. From the
computer as an extension of the brain, McLuhan makes a quantum

leap to electronic technologies as extending mind and simulating consciousness. The polymorphic circulation of imagery between the organic, social, and technological realms suggests the harmonious resolution of conflict. McLuhan's romance with the technology of the
informational society is shared by others and reflected in their language. The biologist Lewis Thomas speculates in *Lives of a Cell* (1974)
that "we are becoming a grid, a circuit around the earth. If we keep
at it, we will become a computer to end all computers, capable of
fusing all the thoughts of the world into a synctium." Bagdikian
proclaims that "news is the peripheral nervous system of the body
politic." Van Rensaeller Potter, another well-known biologist, describes the human being in *Bioethics* (1971), as "an information-processing, decision-making, cybernetic machine whose value systems
are built up by feedback processes from his environment." By as-
cribing the characteristics of our inventions to ourselves, by seeing
ourselves in their image, as these writers do, the distance between
ourselves and those technologies, a distance that is prerequisite to
criticism, is not only reduced: it is eliminated.

By contrast, the essays in this book seek to preserve that critical
distance. They evaluate these myths of information—and others—
both indirectly and directly, and from a wide range of perspectives.
Although the contributors represent many different disciplines—communications and cultural analysis, literary criticism, intellectual history, philosophy, aesthetics, art history, cybernetics, and economics—
their work is best described as cultural studies. The approach is
theoretical and speculative, not empirical or positivistic. Broadly
speaking, the concern is with cultural politics. The tone is neither
exuberantly pro-technology nor shrilly anti-technology, but measured
and open, trenchant and imaginative. Technology is not considered
apart from culture, but rather a part of it, and one part only. Many of
the essays share a marxian framework of analysis. Topics range from
a biological/mathematical model of communication and the mass audiences of the mass media, to the relationship of technology to avant
garde art and an ethics for today's technological society. Although the
essays are grouped together in four sections, the themes of one often
echo those of another.

The equation of information with communication, the identification of information with meaning, are uncritical commonplaces of
contemporary Western culture. What is the relationship between information and communication? between information and cultural
boundaries? between information, noise, and meaning? How do we
distinguish between them? Which do we most value? The essays in

the first section—*Communication, Theory, History, and Cultural Difference*—explore these questions, and many of the essays that follow touch upon them also, as if the boundaries of individual essays were porous.

Our economic system demands that we consume what we produce so as to stimulate further production and thus maintain the strength of the system. The system, however, does not require the health of its component parts to maintain its power. (In this regard the economic system of capitalism is quite unlike the system of an organism—the analogy is misleading.) This insight is the basis of Jean-Pierre Dupuy's critique of the informational society. The endless cycle of the production and consumption of information and services generates less and less meaning, he hypothesizes, because it reduces the possibility of autonomous action within the system. Alienation is the result.

We can read Dupuy's essay as an elliptical commentary on Shannon's mathematical model of communication when it is placed in the context of a social and economic system. To be sure, the meanings associated with the word "information" in these two domains are quite different. In Shannon's theory, information is associated with uncertainty and the resolution of that uncertainty. For Dupuy, and most of the other contributors to this book, information is associated with knowledge, or meaning. Most important for us here is the model of the relationship between signals and meaning, however they are defined. In Shannon's theory, the amount of information conveyed by the message increases as the amount of uncertainty increases. As for what the actual message will be: a message which is one out of twenty possible messages conveys a smaller amount of information than a message which is one out of twenty million possible messages. In popular (not mathematical) terms, this is a fascinating model, because it suggests the pricelessness, the rarity, of a unique message in a world of boundless signals. The analogy to a world of mass-mediated messages, however, is discomfiting: the more signals transmitted, the more precious the single meaningful message will be. In a sense, then, it appears that the way to increase meaning, or value, would be to increase the number of signals transmitted: but is this not a sly tactic of the economic system?

We also can read Dupuy's essay as an argument against the generally accepted identification of information with order and meaning, which view is represented here by the biologist P. B. Medawar: "'Information' connotes order and orderliness or anything that embodies it," Medawar writes in *The Life Sciences: Current Ideas of Biology* (1977). "The connection between information and order is a perfectly straightforward one and reminds us of the fact that the notion of information was imported into biology from communications engineering. . . . 'Noise' connotes the very opposite of information, that is to say randomness or disorderliness." Those who might respond to Medawar have raised the following questions. Is the model of communications engineering that has been introduced into biology appropriate? Heinz von Foerster argues convincingly in his essay that it is not. How is information created? Medawar's model does not account adequately for the *creation* of meaningful information (not just the *reproduction* of information) by an autonomous actor. And what are the political implications of opposing randomness, noise, and disorder to information?

Shannon would argue that the control of noise is crucial, since noise impedes the content and intelligibility of information. Dupuy considers the creative uses of noise instead. Like Daniel Charles, whose essay is on experimental music, Dupuy does not oppose noise to information but regards it as necessary to the *making* of meaningful information, or art, by an autonomous actor. But it is precisely autonomous action that is repressed (or repressively tolerated) by the technological society. In such a complicated system, noise is dangerous, for it threatens a precarious stability (although a little seems necessary for illusions of freedom). Within the framework of Dupuy's thinking, then, both tools for conviviality (technology that furthers decentralization and autonomy of the individual) and much contemporary art promotes the health of the entire globe: they expand opportunities for randomness.

Dupuy is indebted to the research of Heinz von Foerster, whose essay follows his. Drawing on neurophysiology, mathematics, and the theory of Poincaré and Piaget, von Foerster calls our attention to the inadequacy of our widely held models of communication. Information theory should be called signal theory, he remarks, winningly, which would help clear up the common confusion between signal and information. For von Foerster, communication is a mysterious internal process of subtle and reciprocal adjustments of the self to resistances in the world which is not-self. He believes the metaphor of a transparent, frictionless exchange of information is misleading. Von Foerster's "Epistemology of Communication" combines a model of the self-reflexivity of much contemporary thought in the humanities with what we might broadly call ecological thinking. For him, as for Dupuy and Anthony Wilden, whose essay appears in the last section, the potential for creating information is everywhere, but information which has meaning is created only through an act of cognition or as Wilden puts it in *System and Structure* (1972), "information is everywhere, but knowledge can only occur within the ecosystemic context of a goal seeking adaptive system."

If we are bound by our own self-referentiality, we are also culturally bound, as Magoroh Maruyama reminds us. Western communications and transportation technologies may connect disparate points of our planet, but these places are by no means culturally homogeneous. Nor do people who live in the same place necessarily share the same notions of what constitutes information or communication. Maruyama distinguishes between classificational information, which is dominant in the United States, and contextual information, which he illustrates with examples drawn from Japanese culture. His essay is itself an illustration that all communication necessitates context, that without context there is no meaning; and that contexts confer meaning because there is a classification of contexts (Maruyama's charts are an encomium to this classification of contexts). Maruyama does not discern a movement toward planetary cultural homogeneity. Rather he values cultural heterogeneity and calls for an awareness of "poly-epistemological systems," the understanding that people who ostensibly live within the same system often possess different epistemological models of information and communication.

The closing essay in this section, by Eric Leed, approaches the myth of postindustrial society from the point of view of intellectual history. Leed shows how our recent interpretations of Western history in terms of communications revolutions—the shift from oral culture to print culture to electronic or aural culture—are projections of our desire to reconcile the contradictions in modern society between membership and autonomy: the vision of postindustrial society would transcend those contradictions. Like von Foerster and Maruyama who precede him, and Negt and others who follow, he calls into question —but for different reasons—the transmission model of communication that continues to dominate communications research (in the United States at least), revealing that it rests upon the fiction of the autonomous individual which is the hallmark of classical liberalism. But while Dupuy and most of the other contributors to this book are ideologically opposed to the vision of postindustrial society, Leed— and with him David Hall—is willing to entertain some of the possibilities it raises.

The second set of essays—*Technology, Mass Culture, and the Public Sphere*—deals with the relationship of mass-mediated culture to the public sphere. According to McLuhan's myth of the informational society, the global village would be the postindustrial equivalent of the public sphere. Is this credible? If so, is it desirable? Do the mass

media stifle communication? Or does this very question betray the
vested interest of the intellectual who harbors contempt for the masses?
The mass media may have created a mass culture, as Raymond
Williams believes, but does it follow that the mass media have created
the masses—by which we generally mean the unthinking public—as
well?

In these essays, the "public sphere" refers both to the concept
developed by Habermas in *The Structural Transformation of the Public
Sphere* (1962) and the term used by Hannah Arendt in *The Human
Condition* (1958). Habermas draws his concept of the public sphere
from the tradition of the Enlightenment in which it referred to a
space between the private sphere and the state where rational and
critical discussion of art and politics was possible. Arendt also associates the public sphere with a space in which rational discussion
takes place, but her model is the city state of ancient Greece and thus
for her the public sphere did not exist outside the state but was
identical to it. However, both Habermas and Arendt trace the historical decline of the public sphere, and both believe that the mass
media render communication impossible: the stuff of culture has been
transformed by the technology of the mass media, controlled by
capitalistic interests, into commodities.

Following and expanding on Habermas, the first three essays in
this section call for the creation of a counter public sphere and acknowledge the very serious difficulties which this involves. In "Mass
Media: Tools of Domination or Instruments of Liberation," Oskar
Negt argues that the development of a counter public sphere based
on tools of communication other than the mass media—material such
as books and posters produced outside the official publishing industry,
information machines such as tape recorders which can be used privately, and so forth—is ultimately doomed to failure, although they
are useful tactics. Negt's essay moves back and forth between theory
and practice, which is entirely appropriate because he locates the
basic problem in creating a counter public sphere in the dislocation
between theory and practice. The theory of the intelligentsia, Negt
believes, cannot reach the proletariat in practice. To be successful, a
radical protest movement must incorporate the mass media into its
political strategy, even though this presents an apparent contradiction: the mass media themselves are an embodiment of the problem.

Two marxian analyses of the nature of this problem follow. Con-
centrating on popular films—*Snow White and the Seven Dwarfs* (1936)
and *Rocky* (1977)—Jack Zipes shows how the culture industry appro-
priates forms—in this case, the fairy tale—that have traditionally served
to stimulate the imagination and uses them instead to transmit the

ideology of the industry. In "The Political Economy of Social Space," Andrew Feenberg shifts our focus to another technology—transportation. Ironically, one of the results of planetization, he argues, is that people are not brought closer together in a newly constituted public sphere but larger distances are created between us.

The last two essays in this section stand as a kind of riposte to the first three. Neither of them deplores the decline of the public sphere in any conventional sense. David Hall makes use of Eastern thought to construct a model of technological society as an anarchistic utopia. His philosophical and cross-cultural perspective exempts him from the pitfalls of McLuhanism. Hall suggests that the marxian concept of alienation, while appropriate to industrial society, is no longer relevant in emerging postindustrial society. On the contrary, he proposes that the virtue of mass communications and the information machines is that we can in fact alienate ourselves from them, freeing ourselves for pleasure in the expanding private sphere. Hall's attitude resonates with that of John Cage and Murray Schafer, artists whose work is discussed in the next section of this book. For they all counsel us not to take the world too seriously, to rid ourselves of the "high moral seriousness" associated with the Western desire to impose order on a chaos which we fear too much. Just as Maruyama urges us to accept "poly-epistemological systems," so Hall asks us to accept a many-ordered world.

Jean Baudrillard's essay returns us to concerns central to the first section of the book. With Dupuy, he believes that the dissemination of more and more information results in less and less meaning. In metaphors drawn from atomic energy and cybernetics, Baudrillard writes of the irrational violence of the mass media. Bombarding the masses with signs, the mass media do not liberate their energy but render them inert: "it is a gentle *semi*urgy which controls us." But like Hall (although the two could not be more unlike in other respects), Baudrillard is not scandalized by this. The very silence of the masses, their fascination with the bemusing circulation of signs, their indifference to "meaning," is an effective method of resisting the imposition of a particular political point of view. For Baudrillard, it is a mistake to view the mass media as either the tools of liberation or instruments of meaning, an error which proceeds from either regarding the masses with contempt or mystifying them.

For all the authors in this book, the relationship between theory and practice is a critical one: theory implies practice, a cultural politics. The third section—*Art and Technology: The Avant Garde*—considers the nature of a certain kind of practice in twentieth-century Western

society—the making of art. What is the role of the arts in the technological society? the relationship of the avant garde to cultural politics? the relationship of the various media to mass media? If, as for
Harold Adams Innis, the lifeblood of a society is the circulation of
information, do the experimental arts participate in this in a meaningful way? Or should the role of art be precisely to empty itself of
messages? Have the information machines and information theory
transformed the arts? Or is change illusory? The essays in this section,
with one exception, reject the notion that there are fundamental contradictions between art and technology. As they see it, the basic
question is whether technologies can be separated from the technocratic system and *used* differently to challenge that system. The section addresses some of the same issues as the preceding section, but
in a different domain—the arts instead of the mass media, the limits
of the private in the political sphere.

Baudrillard's essay is a link between the first and second sections,
and Andreas Huyssen's essay connects the second and third. Huyssen
concludes that the avant garde cannot recover the historical role it
played in cultural politics in the twenties and thirties and recom-
mends that we turn our attention to the study of mass culture instead.
Technology may have contributed to the development of the historical
avant garde, but the technology of the culture industry was also responsible for its decline, as well as the decline of the public sphere.
Mikel Dufrenne and Daniel Charles are more optimistic. Both believe
that new technologies can be used playfully—that is, appropriately—
by artists. "Torn from the system of domination," Dufrenne writes,
"technique perpetuates the friendship that art reveals between man
and the world." Independent cinema is an example of this. So is John
Cage's work, which both Daniel Charles and I consider. In Charles'
piece, the new music and French deconstructionist theory come together in a McLuhanesque vision of the circulation of the near-pure
signals of music, the most minimally encoded of the arts. Like white
Gaussian noise, which contains all the frequencies equally and is the
epitome of the unexpected and the origin of all sounds, this space is
an absence necessary to the making of new meanings.

But the problem is that even with the best and most poetic of
intentions, artists, like users of the mass media, may reinforce the
ideology of the technocratic society. Or their work may simply exist
on the margins. Is Dufrenne's optimistic vision born out by experience?
As I indicate in my essay, the record is mixed in John Cage's work.
Jack Burnham argues that although electronic technologies were "used"
differently by artists in the sixties and early seventies, their art failed.
Why? Burnham speculates that the ultimate reason is that there is a

fundamental dissimilarity between these two "systems of human semiosis," between the mythic consistency of technology, which embodies our drive to empiricism, and the mythic consistency of art, which has its origin in the mysteries of the human psyche. Contrary to Huyssen and in an almost mythic vein, Burnham concludes that the human strength of art may inhere in its cultural weakness. This is an argument in praise of marginality.

Circulating through *The Myths of Information* are two major propositions regarding cultural theory and action: we must think cybernetically, and we must trust in de-centered action in everyday life. These two mandates are related, for cybernetic thought asserts that the part can never control the whole and that change in a system comes about from shifts in the structural relations, or constraints, of the system. This does not imply a technological determinism. As Wilden points out, the appropriate terms in cybernetic thought are not "free will" and "determinism," but rather "goal seeking" and "restraints," between which there is no necessary contradiction or opposition.

Each of the two essays in the closing section—*Cybernetics, Constraints and Self-Control*—focuses on one of these approaches, but both are object lessons in the interrelation of theory and practice. Wilden's piece is predominantly, and dazzlingly theoretical. Through an historical parable—the model of the medieval cosmos—he quite properly urges us to be precise in our application of cybernetic models imported from engineering and systems thinking to biological, social, and historical systems. This is an important reminder. We must be careful not to conflate cybernetic models into a flabby holism that can serve whatever purposes we have in mind. Wiener's original definition of cybernetics as the science of control and communication included *living* as well as inanimate systems. In the West we seem to have occluded this fact. We commonly associate cybernetics with computer science, information theory, and systems design, depriving cybernetics of its ecological insights.

On theoretical grounds, Wilden opposes consciously-engineered solutions to the crises of capitalism. Change cannot be imposed on the system from the outside, but must come from within (Wilden's use of Chinese models and myths reverberates with David Hall's). Our dilemma is a painfully peculiar one. Of all social systems, it appears that capitalism is the only one whose dominant ideology does not perform the long-range function of survival and adaptation. Thus, to define and apply, consciously, a large-scale solution to our problem is contradictory from a theoretical point of view.

What can we do? Jacques Ellul, whose essay closes the book, speaks directly to this point. We cannot escape the technological society. We cannot claim to be anti-technique. But we must recognize that technique always involves us in questions of power. Ellul advises us to adopt in our everyday life a personal ethics to do less. Like Illich and Schumacher, he urges us to reject the technological imperative, that seductive assumption that because we *can* do something we *must* do it, or inevitably *will* do it, which amounts in the end to the same thing. Ellul's ethics is consonant with the cybernetic theory of Wilden: *"we must choose, a priori, non-intervention each time there is uncertainty about the global and long-term effects of whatever actions are to be undertaken."*

This book, then, is an exploration and critique of some of the mystifications of postindustrial culture defined as the age of communications. It takes a hard look at theories of communication based on exchange models within the framework of a broad theory of culture that includes the mass media as well as art and other works of the social imagination. Finally, the underlying premise of *The Myths of Information* is that any research in contemporary culture makes sense only when it is rooted in a social theory or historical perspective which treats the present as an historical problem.

I
Communications Theory, History, and Cultural Difference

Myths of the Informational Society

JEAN-PIERRE DUPUY

My theme will be that the more we "communicate" the way we do, the more we create a *hellish* world. I take "hell" in its theological sense, i.e., a place which is void of *grace*—the undeserved, unnecessary, surprising, unforeseen. A paradox is at work here: ours is a world about which we pretend to have more and more *information* but which seems to us increasingly devoid of meaning. I will argue, as would Ivan Illich, that this is due to the inevitable and ironical counterproductive effect of the way in which we produce, process, and consume information.

I. The Myths of the Postindustrial Society

I suspect that a double myth of both an economic and political nature looms behind any postulation of a postindustrial society. This double myth must be denounced with determination, for it rests on erroneous analyses and, worse, aims at an ideal which, I believe, is dangerous.

Let me consider, first, the *economic dimension* of this myth. Most of those who speak approvingly of a postindustrial society describe it as a new stage in the evolution of society, an advance in the liberation of mankind from material constraints. The "argument" is this: material needs largely having been satisfied (thanks to the tremendous increase in work-productivity), man at last can turn to the satisfaction of more noble

3

and ethereal wants. These "postindustrial" or "*im*material" needs include health, education, "culture," environmental protection, travel, security, leisure, better relationships with others, and (why not?) happiness. The inputs of industrial economy consist essentially of matter and energy which are transformed by human labor, and its outputs are tangible goods; in contrast, postindustrial economy will be a service economy in which information is both the main input and the main output. They argue further that the well-known inflation of the tertiary sector in Western economies and, even more significant, the tertiarization of employment in the primary and secondary sectors, are signs of the irresistible evolution of our society toward this golden age of services, which Daniel Bell defines as "a game between people, *unmediated by things.*"[1]

It is ironic that American proponents of a postindustrial society have appropriated Marx's theory that man will enter the "reign of liberty" only after escaping the "rule of necessity," that people will finally free themselves from the world of things and commune together in harmony (all obstacles to direct "unmediated" communication having been elim- inated) only after productive forces have reached the peak of their development and bodily needs have been sated, a process begun with the genesis of capitalism. This brings me to the *political dimension* of the myth. The proponents of postindustrial society believe that the growth of information networks and mass communications will bring people and nations together, fostering mutual understanding and peace. They point out that whereas the economic growth of material goods was based on competition, envy, the pursuit of selfish interests, and individualism, the informational society will be an immense harmonious community, a village spanning the whole planet. The ideal is that of a conflict-less, self- transparent society reconciled with itself (here one cannot help recalling the marxian image of a utopian society which is so transparent to itself that, as Lenin put it, a mere cook might well deal with public affairs).

But far from being a stage in the progressive evolution of mankind, postindustrial society—which I will call *the informational society*—is a phase in the history of capitalism coping with its contradictions.[2] Rather than delivering us from material constraints, the informational society intensifies the struggle for survival and strengthens the radical monopoly

---

[1] In 1969 the tertiary sector represented fifty-seven per cent of all occupations in the U.S. and forty-two per cent in France; in 1974 these figures were, respectively, sixty-two per cent and forty-nine per cent. Daniel Bell, *The Coming of Postindustrial Society* (London: Heineman, 1974).

[2] I define capitalism here as that form of économic and social organization prevailing from Los Angeles to Vladivostok via Peking, i.e., an economy based on the accumulation of commodities (see below).

of economic activity over the social and political dimensions of our life (I define economics here as social activity concerned with subsistence). We have grown accustomed to thinking of archaic or traditional societies as "subsistence economies," presuming that they exhaust themselves in organizing survival. Marshall Sahlins has enlightened us on this point: these societies do not dedicate more than three or four hours a day to material necessities.[3] The only society in fact in which the struggle for survival absorbs the better part of our energies and intellect, occupying all available time and space, is industrial society. Instead of fostering harmony among people, then, the new technologies of communication are aggravating alienation, producing a highly unstable, potentially explosive system. The real question is: why has it not yet burst? The informational society is not a large arcadian community but rather a society in which there can be *no* community because its very complication makes it impossible to assign responsibilities or exert authority. To this I will add what the Gulag teaches us: the ideal of a perfectly self-transparent society devoid of conflict and contradictions opens the way to all kinds of totalitarianism.

Moreover, the belief that men will not really be able to communicate with each other until they are freed from the sway of things is untenable. Relationships between people are defined in and through the process of making things together, building a common world: the two cannot be separated. Without a common world created by *making*, no encounter, no human history, no action would be possible, if only because action, like speech, is naturally transient and cannot remain in the memory of men unless actualized by poets and historians. Hence, men cannot communicate unless they have a common world which unites them, but also separates them, just as a table brings people together insofar as it stands *between* them. This metaphor is hardly a metaphor. In *The Death and Life of Great American Cities*, Jane Jacobs shows that what makes American cities so unbearable is that one must live either in complete isolation or in an intolerable promiscuity.[4] The question of the common material world and the question of communication must, then, be posed simultaneously. For it can not be mere chance that industrial capitalism, which does not value personal relationships and works to alienate people from one another, is also ignorant of what it is for men to have a world which they find when they are born, complete by their activity as creators and makers, and leave to their children when they die. Where are the cathedrals of the industrial society?

But there is another reason, perhaps less questionable, that entitles

---

[3] Marshall Sahlins, *Stone Age Economics* (Chicago: Aldine Atherton, 1972).
[4] Jane Jacobs, *The Death and Life of Great American Cities* (New York: Random House, 1961).

us to speak of an informational society. I have in mind the new intellectual tools used by our societies to scan and represent themselves. Due to the spectacular strides made during the last twenty years in biology, the theory of cognition, and the general theory of systems, the standard mechanistic representations of society have become largely obsolete and have been replaced in the human sciences by a flowering of thermo-dynamic, biological, cybernetic, organizational, and informational meta-phors. Many of these were advanced without any epistemological precautions, and I believe they are not worth much. Some, however, seem to be much richer. I will say a few words about them at the end of this paper.

II. The Reification of Relations with Others and the World

The development of a service economy is necessary to capitalism for its own perpetuation; it can only result in increasing alienation. In the past, the prime mover of capitalist development has been *waste*—of raw materials, energy, human space, and time. It is well-known that the threat posed to capitalism by the possible saturation of consumers' material needs was circumvented initially by a policy of intensive innovation and renewal of products. The life-expectancy of most of the "goods" that make up our daily environment has diminished con-siderably in the last thirty years; older products are replaced by new ones which, usually offering nothing new except their names, are more expensive and contain more un-renewable raw materials and energy (it has been estimated, for instance, that eleven billion dollars have been paid *each* year by American drivers because of automobile model changes[5]). But today's energy ecological crises have opened the eyes of the brain trusts of world capitalism. The gentlemen from the Club of Rome and similar organizations which invent the tactics of worldwide industrialization now acknowledge that economic growth can only be sustained on a different basis. Whether they talk of "new growth" or a "new international economic order," the underlying strategy is the same: let's send our heavy industries abroad where they can pollute the countries of the Third World, spoiling *their* landscapes, deadening *their* workers, and disrupting *their* time and space, and let's keep for ourselves the growth of immaterial productions which do not poison the atmos-phere, are suited to decentralized locations, and enable us to solve to a large extent the problem of unemployment.

Since Marx, we have understood that capitalism begins when human labor loses its power to produce *use values* for the worker himself, thus

[5]Franklin M. Fischer, Zvi Griliches, and Carl Kayser, "The Costs of Automobile Model Changes since 1949," *The Journal of Political Economy*, 70 (1962), 433–51.

transforming labor into a commodity which has only an *exchange value* for the worker. The working relation (or, in marxian terms, production relation) is thus *reified:* the worker is turned into a commodity. But the working relation is not man's only relation to other people or to the world. Would it not be possible to transform other relations into commodities as well? to produce them industrially? to sell them? Would this not provide work for the unemployed and favor the pursuit of growth? The possibility of creating these markets clearly cannot be overlooked by capitalism at a time when the energy and ecological crises preclude most of the traditional avenues of economic development. In fact, the massive development of services has no other grounds than these.

In order to understand the effect of capitalism's appropriation of the delivery of services, we can apply the concepts of *use value* and *commodity,* notions drawn from classical economics, to areas traditionally considered foreign to them. Let us examine (as Jane Jacobs prompts us to do in *The Death and Life of Great American Cities*) the function served by street crowds, those people whom we may scornfully define as "non-productive," but thanks to whom, the city, between 9 a.m. and 5 p.m., is not quite a desert and all residential districts not yet death-traps. The idle stroller whose aimless presence on the sidewalk contributes to the safety of the city or the old lady sitting behind her window, "giving eyes to the street"—these people, without knowing it, are productive of values such as safety, liveliness, and conviviality. We must also recognize that these services could be produced in other ways. In the "eyeless" streets mentioned by Jacobs, safety, for example, could be enforced by private militia troops or bureaucratic police officials.

The example may be generalized. Any use value can be produced by bringing into play one of two basic modes of production—an autonomous mode of production or a heteronomous mode of production.[6] We can learn, for instance, by becoming acquainted with the things of life inside a meaningful environment, or we can learn within the walls of a professional institution. We can remain healthy by leading a wholesome life, observing hygienic principles (an "art de vivre"), or we can be taken care of by a score of therapists. We can have an active relationship with the space in which we live based on movements, such as walking or cycling, that feed on metabolic energy, or we can shift about in all directions, seeking to leave our place as often as possible transported by motor vehicles industrially produced and regulated. We can render a service to someone asking for help, or we can refer him to people who provide those services. In contrast to the heteronomous mode, what is produced by the

---

[6] These concepts have been developed by Ivan Illich in *Medical Nemesis: The Expropriation of Health* (New York: Random House, 1976) and Jean-Pierre Dupuy and Jean Robert, *La Trahison de l'opulence* (Paris: Presses Universitaires de France, 1976).

autonomous mode cannot in general be measured, estimated, compared with, or added to other values. The *use values* produced autonomously thereby escape the control of economists or national accountants.

I do not contend that heteronomy is hell and autonomy paradise. The autonomous capacity to produce use values may be significantly invigorated by the heteronomous mode of production. The professional therapist, antibiotics, or the predetermined curriculum may, to a certain extent, improve what people could do on their own or through mutual aid. This is too obvious to be emphasized. Yet my hypothesis is that beyond certain critical thresholds, the heteronomous production of values generates such a thorough rearrangement of the physical, institutional, and symbolic environment that the autonomous capacity to produce use values is paralyzed. The paradoxical effect of such a restructuring is that the more pervasive the means brought into play by the great heteronomous service-producing institutions, the more they become an obstacle to achieving the aims they are supposed to serve: medicine makes people sick, schooling makes them dull, transportation systems immobilize them, and communication systems make them deaf-and-dumb. The service economy has gone far beyond these critical thresholds. Relations with other people, the world, or oneself, which we call education, habitation, or health but confuse with the educational system, urbanization, or the medical establishment, become things or substances or commodities whose degradation our social engineers, in spite of their endeavors, seem quite unable to prevent.

Let us consider in a bit more detail the case of medicine. It is a well-known fact that in our *over*industrialized nations doctors' consulting rooms are crowded with strikers. Since "free" countries recognize the right to strike but *only* with regard to work, going on strike as a husband, lover, father, son, teacher, pupil, or leader is not tolerated. Being sick is one of the rare forms of deviance that is actually accepted. Thus it is socially permissable for any problem of ill-being (bad relations in work or married life, lagging behind at school, and so on) to result in an assistance claim addressed to the medical institution, a claim which as a rule is not presented in its actual terms but is more or less disguised with somatic terms. In such a context, the proliferation of medical services multiplies circumstances in which people are told that they are ill because something *in them* is out of order, rather than because they cannot adapt themselves to an environment that is perhaps intolerable. A French drug-producing laboratory recently advertised a medicine to treat the complaints that arise from "new urban developments," or, as one would say in the United States, the "mid-Manhattan disease." This is only one example among many. The medicalization of life and society both testifies to and causes a loss in people's ability to settle their own problems within personal networks. The capacity of individuals to refuse the unacceptable is thus weakened and their resignation from social struggles reinforced. Medicine is becoming the alibi of a pathogenic society.[7]

Ivan Illich has shown that the unbounded development of medical services has been destructive of another fundamental human capacity—the power to cope in a conscious autonomous way with a number of intimate threats such as pain, handicaps, or death. The histories of pain, illness, and the image of death reveal that men have always managed to cope with these threats by giving them *meaning*, interpreting them in terms of what anthropologists call "culture." Today, however, the expansion of the medical establishment is closely connected with the spreading of the "myth" that the elimination of pain and the indefinite postponement of death are not only desirable objectives but that they also can be attained with the increased development of the medical system. A problematic arises: how do we give meaning to something that we seek to eliminate by all means? This issue is linked up with a more general characteristic of industrial—or postindustrial—societies, namely, that entire aspects of the human condition are becoming meaningless. This is *both* the result and the mover of the great myth-carrying heteronomous systems that keep growing and keep telling us: "don't bother about these meaningless spaces, *we shall eliminate them.*"

The space and time that we live in and construct will stand as a final illustration. The French anthropologist Leroi-Gourhan has shown that in all archaic and traditional societies, the space that is experienced has a property known as "connexity": any two points in that space can always be connected by a continuous path without people having to get out of the space.[8] The first society in history to have disrupted this connexity is the capitalist society, in which the personal space of each person has burst into distinct and distant centers—his dwelling, his working place, a few symbolic city spots, shopping centers, and the mythical "elsewhere" of leisure and escapism. Between these centers is nothing but meaningless wilderness that one seeks to cross as quickly as possible by *entrusting* ourselves to the transportation system. We protect ourselves, for example, from the space of the highway by metallic bubbles which often are transmogrified into coffins. Everybody is familiar with the recurring news item describing dozens of drivers who ignore a wounded man

[7] On all these points see Jean-Pierre Dupuy and Serge Karsenty, *L'Invasion pharmaceutique*, 2nd ed. (Paris: Editions du Seuil, 1977) and Ivan Illich, *Medical Nemesis*.

[8] To learn about the "experienced space" of an individual, we may ask him to draw it and then compare the drawing with an "objective" map of the place. On this point, see Alain Metton and Michel-Jean Bertrand, "Les Espaces vécus dans une grande agglomération," *L'Espace géographique*, 3 (1974), 137–46; André Leroi-Gourhan, *Le Geste et la parole* (Paris: Albin-Michel, 1964).

dying on the roadway. Monsters! you will say. But studies have been
made that permit a different explanation. For the driver, images reaching
him from outside his vehicle are signs that a mere stepping on the gas can
obliterate, just as a person watching television can get rid of war atrocities
by turning the off button. This, of course, is not an accidental comparison.
The ways in which a "transported" person and a viewer perceive the
world have "falseness" in common, not because they are made up of
intrinsically spurious information, but because they place the spectator in
an illusory position of omnipotence with regard to a representation of the
world which appears or vanishes from sight at his own sweet will.

It is on this particular structuring of space—and time—that the myth
of transportation systems is grounded: the myth of coalescence, a return
to one's traditional neighborhood through the ideal of *cancelling* this
meaningless kind of space-time. A Swiss airline company regularly buys
two-page spreads in the main European weeklies in order to display the
picture of a traditional and apparently magnificent city with its buildings,
squares, canals, and rivers. But, if you take a closer look, you will notice
that this city is a monster: it is a hybrid of the finest districts of the finest
cities in Europe. Only a river separates the Red Square from Place de la
Concorde, and Via Veneto leads directly into Picadilly Circus. And the
caption reads: "thanks to Swissair, Europe is boiled down to just one
city!"

While it is true to a degree that the whole world is present to the jet-
set wherever there is an airport and a host of slaves wasting their time—in
order to save the passenger's time—in a maze of slow ways and urban
traffic jams, for the majority of people, this kind of retrieved "global
village" connexity is much too loose a web: they either slip through the
meshes or get trapped in the knots. To them the city is a fragmented,
inaccessible entity, and roads separate more people than they connect.
For the ubiquity promised by transportation systems conceals with
difficulty the dreariness of real facts. The average inhabitant in our cities
must waste up to five hours a day simply to *move* in industrial space,
sometimes immobilized in his daily orbit, sometimes absorbed in a
stultifying job that will supply him with the means necessary for his own
commuting.

III. More and More Information, Less and Less Meaning

> We know about the screech of the owl,
> we know everything about the owl, more
> than the owl itself: so anyhow it may well
> keep hooting at night!
>
> René Victor Pilhes, *The Accuser*

In all their relations with others and the world, people in postindustrial society are becoming *consumers* of *things* produced for them by institutions. Their capacity to *act* is atrophied. This replacement of *acting* by *making* or *fabricating* (and *consuming*) may be construed as the incarnation in everyday experience of the deep-rooted distrust of action and its attendant risks in Western thought. I am thinking, for example, of Hannah Arendt's wonderful work on this subject.[9] Acting is the human power to beget histories, to set processes in action through the network of human relations. It requires courage because, after acting, no one can utterly control the ramifications of his original act: the histories born from the collision of people's individual actions and reactions are the locus of surprise, of random, unforeseeable, improbable events. These are histories actually *made* by nobody and whose authorship no one is entitled to claim. But for those who watch them they are histories that can make *sense* in that they have both meaning and direction. This mysterious harmony willed by no hidden string-puller, but in which people recognize themselves, has usually been interpreted in terms of *making* or *fabricating*: what is looked for is the author backstage. It is the "unseen hand" mentioned by Adam Smith, or the "mover" of history, or the Marxian notion of the "making of history" by the proletariat.

Hannah Arendt was one of the first to advance a radical critique of totalitarianism and philosophies that pretend, once and for all, to define the sense of man's adventure. She disclosed their secret ambition: to render human history an artifact. In principle a radical solution to the problem raised by the unforeseen nature of action, but at what cost! For we do not make sense the way we make a table. The substitution of managerial techniques for politics and ethics (understood as the art of counteracting the risks of action) inevitably results in the negation of meaning or sense. The composition of individual behaviors then produces collective histories from which the real actors are alienated. They become the spectators of their history and feel as if they were the playthings of natural forces external to the web of human relations.

It is particularly noteworthy that Hannah Arendt centered her political philosophy around the concept of *autonomous action productive of radical novelty* which reappears (as a simple analogy or, perhaps, as more than that?) in the critical thought that originated in the crisis of determinism in certain areas of science, particularly in biology and the human sciences. How can we, in these fields, account for the amazing increase in complexity or the emergence of radical novelty that, for instance, characterizes not only such phenomena as non-directed learning, the development of organisms, and the evolution of species,

[9]See in particular Hannah Arendt's *The Human Condition* (Chicago: Univ. of Chicago Press, 1958).

but also artistic and literary creation, the appearance of new cultures, new human relations, and new life-styles?

One may naturally deny that there exists in the world any such thing as creation or the emergence of new meanings. Biologists who persistently refer to a so-called genetic program deny it: everything would result from the development of a pre-written program. Anyone who argues that *Hamlet* is nothing but a particular combination of known and limited signs also denies it; they concede it is quite unlikely that an ape could type and reproduce the text, but maintain it is not altogether unthinkable. Jorge-Luis Borges has demonstrated the absurdity of such views by carrying this logic to its utmost conclusion. In "Pierre Ménard, Author of the Quixote," Borges indulges in one of his favorite games: he describes in every detail the imaginary work of a no less imaginary writer.[10] The latter, an unknown author living in the south of France at the turn of the century, was of special interest to Borges because he was engaged in a word-for-word, line-for-line, very complex re-writing of Miguel de Cervantes's work. When he died—most unexpectedly—he had only been able to reproduce chapters IX and XXXVIII and a fragment of chapter XXII. Borges-as-narrator makes a comparison between the two texts, which of course—since this is the basic hypothesis—are perfectly identical. He easily demonstrates the greater subtlety but also the greater affectedness of Ménard's text as compared to that of his illustrious predecessor: the context is totally different, many events have taken place during the three centuries separating the two works, including those brought about by the publication of the *Quixote*.

I will not insist on how Borges's ingenuity makes the best of that situation. The message is clear enough: the meaning of a work is not intrinsic to it; it cannot be exhausted by the particular combination of words and symbols that constitutes it. But now let us suppose we had a *perfect* knowledge of the determinants of the meaning located in the *context* of the work, so that meaning would be a univocal function not of the text but of the whole: of both the text and context. Paradoxically, this knowledge would abolish precisely what we are seeking to account for: the emergence of a new meaning.

The same Borges is the author of the well known "Library of Babel," in which he describes a library-universe made up of librarians and books, the latter all having the same number of pages and of signs per page, the signs amounting to twenty-eight.[11] It is through the content of these books that people seek access to the meaning of their world. Unfortunately, most of the books have no meaning whatever in any known

[10]Jorge-Luis Borges, *Obras completas* (Buenos Aires: Emecé Editores, 1974), pp. 444—50.
[11]Borges, pp. 465—71. The twenty-five letters of the alphabet, two punctuation marks, the space between the words.

language. But sometimes, in some of the books, emerges a sentence, or two lines, or a whole page that seems to convey a comprehensible message. Over the course of centuries a religion has developed which asserts that *all* books exist—in very large but not infinite numbers—and that there must be one that sums up all the others and contains the meaning of the universe. Some people say this book has already been discovered, but was destroyed because the secrets it revealed were too horrible. Borges-as-narrator logically concludes that this destruction is not irremediable as there must exist millions of nearly perfect facsimiles differing from the right book by just one word, one letter or one comma, just as there must exist refutations of this book by the thousands. And the very piece Borges is writing must also *be* somewhere inside this incredible universe. . . .

Let me interpret: a world without constraints, without order, a world in which everything would be possible, would have no meaning. But in a perfectly deterministic world whose determinations would be perfectly known, the creation of meaning would be just as unthinkable. There can be no creation for an omniscient God. It is the epistemological position of the observer perceiving a somewhat—but not totally—ordered world that gives rise to the feeling that there exist autonomous systems capable of creating radical novelty. This, indeed, was the conclusion reached by biologists, at least by the most advanced of them. The self-organizing capacity of living creatures (or what seems to us to be so) results from their ability to cope with random aggressions (or what seems to us to be so) through a sequence of disorganizations and reorganizations at a higher level of complexity. Creation feeds on disorder, on chaos. Randomness is part and parcel of organization. Disorder is inherent in the definition of order.

It was Heinz von Foerster's merit to have clarified and formalized these ideas for the first time under the principle of "order from noise." [12] The talented French biologist Henri Atlan has reworked and developed them, suggesting a new formalization capable of surmounting the two standard limitations of Shannon's information theory: namely, that it can account neither for the *creation of information* nor for the *significance* of information. [13] All I can do here is to describe his approach in a few words. Imagine a self-organizing system that we observe and about which we do

---

[12] Heinz von Foerster, "On Self-Organizing Systems and Their Environments," in *Self-Organizing Systems*, ed. Marshall C. Yovits and Scott Cameron (New York: Pergamon Press, Symposium Publications Division, 1960), pp. 31–50.

[13] Atlan has replaced the notion of order used by von Foerster—repetitive order, i.e., redundancy—with that of complexity in the sense of absent information, i.e., the entropy of information. See Henri Atlan, *L'Organisation biologique et la théorie de l'information* (Paris: Hermann, 1972).

not know everything.[14] Shannon's theory nevertheless enables us to quantify the *absent information* we need to make a complete description of the system. But it is meaningless information. Suppose the system happens to be disturbed by such or such an event; for us, it is noise, but it is obviously meaningful for the system since hypothetically the latter does react by changing its structure. The subsequent increase in complexity will result, for us, in an increment of that *meaningless* information measuring the *absent* information: it is the projected shadow that will point out to us the emergence of new meanings. And it is through this double negation that Shannon's theory may account for them.

It is essential to understand that the level of information which escapes the observer is largely left to his own discretion. There are people who are never surprised by anything and never learn anything because they have crude ideas in all matters and/or schematic views fixed once and for all. Naturally the properties of the observed object must come into play: if the object is an artifact made by myself, its complexity, in terms of the information that escapes me, will amount to nil (even if its complication, estimated for instance by the minimum number of stages necessary for a Turing machine to describe it, is very high). The more constraints I impose on a system, the smaller this missing information will be.

Heinz von Foerster, Ross Ashby, and their team are famous for having designed and made automata that are able to *surprise* those who observe them—a property on which the richness of life is actually grounded. By automaton—we also say "machine"—I mean not so much a combination of articulated parts as a mathematical being that may happen to be put in concrete form as a mechanical or electrical system. A stimuli-response system, characterized by a well-defined input-output relationship, constitutes such a machine, but one which is called *trivial:* its behavior can by no means surprise us. A non-trivial machine is one which consists of several input-output relationships, with the passage from one to the other taking place according to a certain determinism (for example, by bringing into play the whole past history of the machine). Although deterministic, such a system—if we are ignorant of its determinism—will seem to us quite unforeseeable in its effects and endowed with autonomy. This, after all, corresponds to the definition of an "automaton." According to the quantity of information that will measure our ignorance about it, we will say that the machine is, for us, more or less trivial.

Here too, it is necessary to understand that the property of being more or less trivial is not intrinsic to the observed object but is a

---

[14]As the foregoing shows, if we knew it completely, it could not appear to us as self-organizing.

characteristic of the representation made by the observer himself. For instance, in the case of a person used to "thinking" in terms of simple relations of the following type: American = cowboy, Russian = vodka drinker, French = fornicator, it is most unlikely that the scanty attention he pays to others will enable him to consider them as anything more than trivial machines.

(I will make a short digression here to illustrate these notions through a personal example. The obstacle of language prevents me from being autonomous with regard to my written paper; all I can do is read it, without any variations or fanciful embroidery. I am thus behaving as a trivial machine: what is orally expressed by me is a univocal function of what is written. You will note that I also had to write down this last sentence. But also this last one. And also that one, and so on. This should not be regarded as a joke but as an illustration of a more difficult concept connected with autonomy, one that has been formalized by such researchers as Heinz von Foerster and Francisco Varela and that bears on *self-referential propositions* and their relationship to infinite recursions.[15] In the present case, it stands to reason that if I want through my discourse to rigorously establish the following proposition bearing on my own discourse, namely that it is not autonomous, then I must never stop. Because if I stop, I will either have said something that is not written—but that is impossible in principle—or I will have read something about which I would have failed to say that I had had to write it. My lack of autonomy forbids me to speak of my lack of autonomy, at least within a finite given time.)

I can now introduce a conjecture made by Heinz von Foerster that will take us back directly to the central paradox of the informational society. I will state it in very approximate, intuitive, and anthropomorphic terms. Let's take a certain number of automata and let's couple them in a certain way: for instance, the output of one will be injected as the input of another. The conjecture is that the more trivial the individual machines (we had better say, the more trivial each one's representation of the other machines), the less possible it will be for them to control the global system they constitute, with this result being more likely as the number of machines is increased or as they are more interconnected. These characteristics (large number of elements, high interconnection, triviality of the connections) define the *complication* of a system, which

[15]See Heinz von Foerster, "Objects: Tokens for Eigen-Behaviors, for Jean Piaget on his 80th Birthday, July 2, 1976," in *Hommage à Jean Piaget*, ed. B. Inhelder et al. (Neuchatel, Switzerland: Delachaux and Niestle, 1977) and Francisco Varela, "A Calculus for Self-Reference," *Int. J. General Systems*, 2 (1975), 5–24.

must not be confused with the notion of complexity. The absence of control affects both the power to act and the power to understand. As regards behaviors, it results in what may be called a structural instability of the system, i.e., an inherent incapability to cope with random disturbances, or noise, affecting its internal connections. As regards cognition, it results in a blocking of the flow of meaning between the individual elements and the whole they make up: they do not "recognize" themselves in what nevertheless originates from the synergism of their behaviors. A certain amount of indetermination seems to be a condition of the capacity of a complex system to regulate itself, to integrate the disturbances that affect it, and to transform them into significant experiences.

Regarding postindustrial society, I suggested earlier the following aphorism: "more and more information, less and less meaning." The paradox may now be elucidated. "More and more information" should be understood as meaning "*less and less absent information.*" Such, indeed, is the ideal of a society which, as it turns interpersonal relations into commodities and thus paralyzes its members' autonomy, tends to look like a beautiful mechanism whose works would be accessible to our total knowledge.

Postindustrial society is characterized by a growing and *trivial* (in the sense of rigid) interdependence of all our acts, whatever their nature or their author or their field of application or the time and place of their occurrence. I cannot move my little finger without the course of the world being affected by it. Everybody agrees, but the proponents of world-wide postindustrialization jump to the conclusion that public life will be more and more communal. Taking my stand on the foregoing, I contend that what happens is exactly the reverse, for at least two reasons.

The first requirement of any communality of life is that each person must be responsible for his own acts, including their unforeseeable unwanted consequences. When social reality seems to result from the action of forces similar to those which generate processes in the natural world, investigating responsibilities becomes purposeless and leads to mythical "whodunits." Conflicts are no longer face-to-face; individuals or groups are no longer in direct confrontation with one another. Consider, for instance, what has become of territorial strifes: the poor are no longer expropriated by the rich but rather through the automatic action of the real estate market. A new form of domination is appearing in which man is no longer dominated by man but in which men are managed by another anonymous collective entity: by *mechanisms*. Theses that used to be regarded as absurd are increasingly accepted as legitimate. [16]

---

[16]See Julien Freund's analysis in *L'Essence du politique* (Paris: Sirey, 1965).

Since no one is responsible, everybody must be. Just as for Jaspers the whole German people were responsible for Nazism and its misdeeds, it is currently contended that "each one of us is responsible for an occasional famine in China" or in Biafra. Meeting the claims of workers at one point of the planet is "responsible" for unemployment at the antipodes. As impenitent motorists, we are all responsible for the black tides that kill our oceans.

In addition, the industrialization and postindustrialization of the planet breed, on a world-wide scale, a highly complicated and trivialized system whose extreme fragility I have mentioned. As its organization leaves little room for the "noise" resulting from the creative autonomy of its components, it becomes more vulnerable to that noise. Taylor's ideal is coming true: in every situation there is "one best way." And it is becoming more and more hazardous to deviate from that way. Consider the ever more rigid and ever more unstable "equilibrium" of nuclear "terror." Consider the fact that the survival of our "wealthy" societies depends more and more crucially on the stability of societies located thousands of miles away. It would be difficult to imagine a more potentially violent situation. Obviously our only way of coping with it is to jump into the vicious circle of programming and vulnerability: more programming entails more vulnerability, which then necessitates and legitimizes increased programming. And, therefore, a heightened concentration of power. The risk of an explosion is real, but there is an even more alarming danger: that we should forget that our main wealth is in man, in his power to marvel and in his capacity to surprise.

Epistemology of Communication

HEINZ VON FOERSTER

As the inquiry into communication progressed at the Center's Symposium on Technology and the Public Sphere, I had the growing impression of having stumbled into a welcome conspiracy to debunk some of our culture's cherished concepts which, on closer scrutiny, appear to be pollutants in the linguistic domain. If we did indeed live in a society in which our ideas and experiences were expressed *through* language, then these pollutants would be minor aberrations, blemishes easily eradicated, and the problem they pose would be unimportant and irrelevant. In our society, however, our experiences and ideas are not expressed *through* language; rather, we have accepted language as *controlling* our ideas and experiences. And in this context a mere linguistic irritant may become a pathogen in the social fabric: I will examine some of these pathogenic linguistic pollutants. I have chosen "communication" as my topic because the very concept of communication, and the assumptions about the underlying principles which make communication possible, are the crucial give-aways in our idea of a postindustrial society, or, indeed, any kind of society where, by the very communicative act, everyone becomes a neighbor. It is this notion of communication which provides the conceptual basis for the structural coupling that allows us to become, theoretically at least, active and congenial neighbors.[1]

My procedure will be, first, to discuss some of the misconceptions that have crept into almost all domains of communication theory, be they rooted in linguistics, semantics, sociology, physiology, mathematics, etc. In most of these domains, metaphors are used which, if not false, are at least misleading, and which in some instances may even turn out to be politically dangerous. Secondly, I would like to report on some of the work done by my colleagues and myself which is relevant to this topic.

Let me begin by presenting some of the notions about communication which are well known, and *therefore*, require careful examination. If you look up "communication" in any dictionary, you find that it means an *"exchange* of information." This notion of communication as an "exchange" rests on the image of a tube: you plop something into one end of the tube, it goes through, and you extract it from the other end. By reversing the process, i.e., by pushing something through the tube from the other direction, the image of communicating is created. The entire process is called the "exchange of information." Circulating through these tubes you may have water, gasoline, or, in some cases, information. In this context, information is considered a commodity, a substance which can be passed through tubes. Moreover, such a substance, we assume, can also run over wires, for we "know" that information, traveling over wires, is passed from one end of a continent to the other. In every textbook of communication theory you will find beautiful pictures based on this image: two little boxes (the transmitter and the receiver) connected by a line (the communication channel). It all seems so obvious.

What is traveling on that wire, however, is not information, but *signals*. Nevertheless, since we think we know what information is, we believe we can compress it, process it, chop it up. We believe information can even be stored and then, later on, retrieved: witness the library, which is commonly regarded as an information storage and retrieval system. In this, however, we are mistaken. A library may store books, microfiches, documents, films, slides, and catalogues, but it cannot store information. One can turn a library upside down: no information will come out. The only way one can obtain information from a library is to *look* at those books, microfiches, documents, slides, etc. One might as well speak of a garage as a storage and retrieval system for transportation. In both instances a potential vehicle (for transportation or for information) is confused with the thing it does only when someone makes it do it. *Someone* has to do it. *It* does not do anything.

---

[1] H. R. Maturana, *Biology of Cognition*, BCL Report No. 9.0 (Urbana: Biological Computer Library, Univ. of Illinois, 1970).

Similarly, lexical definitions of dialogue and conversation also rest on the metaphor of exchange, either the exchange of ideas and opinions or of thoughts and feelings. This suggests that if we are at opposite ends of a dialogue and have successfully exchanged our opinions, I have your opinions and you have mine! Presto! How, for heaven's sake, by the mere vocal noises of the participants in that dialogue this mysterious conversion took place, still remains a miracle.

Such lexical definitions, then, do not get at the root of the problem of communication. We must turn elsewhere, to sociology, perhaps. This field seems like a promising source for a solution to our problem because most sociologists, I venture, would agree that communication is the glue which transforms a mere collection of individuals—an "ensemble of independent elements" as one would say in thermodynamics—into a "society," i.e., into a coherent whole.

The difficulty is that it is not quite clear how this glue works. In order to find out one may be tempted to conceptually break up the system (i.e., "society") into parts and see whether they can be understood. If not, one may continue the process of getting ever smaller parts until a unit is encountered that one feels can be understood. This process is called the reductionist method, and its charm is that, by its nature, it will always succeed! But, alas, the trouble begins when one wants to reconstruct the whole from the parts: the glue being removed, the pieces fall apart. Instead of attending to the parts one should have attended to the glue.

If sociologists can not really help us, perhaps engineers and mathematicians have the answers. Because of the plethora of books with such titles as *The Mathematical Theory of Communication*,[2] *Science and Information Theory*,[3] and *Information Theory and Esthetic Perception*,[4] we are led to believe that these thinkers understand information and communication. However, when we look more closely at these theories, it becomes transparently clear that they are not really concerned with information and communication but rather with signals and the reliable transmission of signals over unreliable channels (this in itself is a formidable task and as we know, fabulous things have been achieved in this domain). How such brilliant thinkers, who created this novel science which is deceptively called "information theory," could confuse two concepts of such profound semantic difference as "signal" and "information," is difficult to grasp unless we remember the historical context of the development of this theory: these concepts, together with those of the

---

[2] C. E. Shannon and W. Weaver, *The Mathematical Theory of Communication* (Urbana: Univ. of Illinois Press, 1949).

[3] L. Brillouin, *Science and Information Theory* (New York: Academic Press, 1956).

[4] Abraham Moles, *Information Theory and Esthetic Perception* (Urbana: Univ. of Illinois Press, 1966).

general purpose computer, evolved during World War II. And during wartime a particular mode of language—the imperative, or the command—tends to predominate over others (the descriptive, the interrogative, the exclamatory, etc.). In the command mode it is assumed that the following takes place: a command is uttered, it reaches a recipient, and the recipient carries out the command. For example, if I shout "ATTENTION!" everyone will come to attention. As Jean-Pierre Dupuy has observed, the command mode can exist only in a "trivial" system, one in which it is assumed that output is uniquely determined by input, in this case, by the command.[5] Thus by analogy with unfailing obedience to commands (signals), it appeared to these thinkers as if "signal" and "information" were identical, for all the interpretive processes (understanding, meaning, etc.) were invisible. The distinction between signal and information becomes apparent, however, when a command is *not* followed, when there is disobedience. Consider again the command "ATTENTION!" but in this case, instead of coming to attention, one young man flashes an obscene gesture and walks away; for him, that command was the signal to take on the responsibility either to become the leader of a revolution or to be shot as a renegade (or in a different context, to be burned as a heretic—note that the Latin *renegare* means "to deny" and the Greek *hairesis* means "choice"). Thus we see that information is *created* when a choice is made, but to be able to choose, one must be free to be a revolutionary. Therefore, an epistemology is a political issue.

To recapitulate: a system in which commands function smoothly is one in which information and signal are indistinguishable from each other. This is the behaviorist ideal. The system is threatened the moment someone behaves not as he "ought" to, but as he might wish to, thereby creating a climate in which the "new" might be born.

If these glances at sociology and the engineering sciences have served mainly to demonstrate that their notions of information and communication are woefully inadequate, they have also given us some clues as to where to look next. For if indeed we have to abandon the notion of a commodity called "information" which changes hands in a process called "communication," then we must also abandon a strategy that keeps us searching among things *outside* of us and adopt one that allows us to look for processes *within* ourselves.

Thus, I turn to neurophysiology, and once again my first task is to debunk one of the most cherished misconceptions about the functioning of the nervous system, namely, that the senses (visual, auditory, gustatory systems, etc.) are responsible for allowing us to see, hear, and

[5]See Jean-Pierre Dupuy, "Myths of the Informational Society," in this volume.

taste the world as it is in its glorious variety and abundance. This misconception is revealed by a fundamental principle—the Principle of Undifferentiated Encoding—which applies not only to the activity of peripheral receptor cells (the cones and rods in the retina, the hair cells along the basilar membrane in the inner ear, and so forth), but to the activity of *all* the nerve cells in our bodies. (One thing even more impressive than the simplicity of this principle is that although it has been well known for more than a century, its enormous significance for a theory of cognition was discovered just a few years ago.) In its most concise formulation, this principle states:

> *The response of a nerve cell encodes only the magnitude of its perturbation and not the physical nature of the perturbing agent.*

Since the response of all nerve cells is a train of electric pulses which follow each other at various intervals (with lower and higher frequencies), this principle states, in other words, that it is *only* the strength of the stimulus applied to a particular cell which determines (or is encoded in) the frequency of the pulse train leaving the cell and eventually interacting with other cells deeper in the brain.[6] This pulse train, however, makes no reference whatsoever to the physical cause that activated the nerve cell in the first place: a blow to your head and the absorption of electromagnetic radiation in the retina produce the same sensation—light!

Of course, different types of receptor cells vary considerably in their sensitivity to different physical agents. For instance, a touch receptor almost completely ignores changes in the molecular mean kinetic energy (usually referred to as "temperature") of its surrounding tissue but will faithfully report a tick marching over the skin, while a heat receptor will be oblivious to the tick, but will respond to the changes in the molecular mean kinetic energy of its surrounding tissue. We must note, however, that these distinctions between cell types can only be made by an outside observer, an exploring neurophysiologist, for example, who can look at the same time at two things: on the one hand, for instance, at a thermometer, and on the other hand, at the electrical activity of a receptor cell which he monitors with a minute electrode that is inserted into this cell. When he sees changes in the cell activity due to changing temperature, he says, "this is a heat receptor"; when there is almost no change in its activity, he may try another cell or another stimulus.

The subject on which this experiment is conducted, however, experiences the change of temperature without the benefit of looking at a thermometer, or changes in brightness without looking at a photometer.

---

[6]Radio buffs may appreciate the fact that frequency modulation is used to its full advantage in the transmission of signals within living organisms.

Her or his nervous system "computes" distinct sensations based on the flow of activity of more than fifty million afferent fibers which all say "so and so much on this place of your body," but do not say what it is. The problem of how indeed the nervous system creates the richness of sensations we experience is not trivial at all! And this becomes particularly clear when I point out that Henri Poincaré, the French mathematician, physicist, astronomer, and philosopher, proved around the turn of the century that with all the signals that come from our sensory apparatus alone, we (our—or for that matter, any other— nervous system) cannot in principle construct (compute) our notions of space, objects, and shapes, that is, our perceptions: sensation is necessary but not sufficient for perceiving. Perhaps, because of this counterintuitive and outrageous proposition (what else are our senses good for if not to give us a picture of the world?), Poincaré's essay entitled "L'Espace et la géométrie"[7] fell into oblivion and, as was true with the Principle of Undifferentiated Encoding, its significance was discovered only very recently.

While the negative part of Poincaré's thesis (the insufficiency proof) is in itself a considerable achievement, it is the positive part, the development of the necessary *and* sufficient conditions for the perception of space, objects, and so on, which makes it so important for us today. What is the missing link? The missing link is *movement*. Poincaré shows that the construction of perceptions is contingent upon the process of changing one's sensations by moving one's body and correlating these changes in sensation with these voluntary movements. If we call the "sensorium" the totality of the faculties of perception, orientation, memory, etc. (as distinct from those of reasoning, volition, affectivity, etc.), and if we call the "motorium" the totality of the faculties of voluntary controlled movement, including walking, talking, and writing, etc. (as distinct from those of peristalsis, pupillary dilation, etc.), then Poincaré's thesis can be restated as:

> The motorium provides the interpretation for the sensorium and the sensorium provides the interpretation for the motorium.

This sounds very much like a circular argument, a *circulus vitiosus*, which according to orthodox logic is cause for logical mischief and hence should be barred from scientific discourse. Such an operative recursive loop, however, is not a vicious circle, it is instead a *creative circle*.

Perhaps the following mathematical demonstration will make it clearer that circular causality does indeed allow us "to ascend" or, as I

---

[7]Henri Poincaré, "L'Espace et la géometrie," *Revue de Métaphysique et de Morale*, 3 (1895), 631–46.

would translate the Aristotelian *anabasis eios allos genos,* "to transcend into another domain." First, consider the primary motor action (behavior) m_0 in which the full range of an organism's movements is constrained in some way (for the moment I will call these constraints "objections" to the organism's intended movements). Next, let s_0 be the activity in the sensorium generated through a function M by the motor action m_0:

$$s_0 = M (m_0).$$

Similarly, let the motor action m_1 which is initiated by the sensorium be a function S of its activity s_0:

$$m_1 = S (s_0).$$

Thus, if we substitute the above expression for s_0, we have:

$$m_1 = S (M (m_0)) = SM (m_0).$$

If we call SM the combined sensorimotor operator, or Op, then the above, rewritten, is:

$$m_1 = Op (m_0).$$

By passing through the sensorimotor loop, this activity will create new activity, which we can write as:

$$m_2 = Op (m_1),$$

or, substituting the expression for m_1 given above:

$$m_2 = Op (Op (m_0)).$$

And clearly:

$$m_3 = Op (Op (Op (m_0)))$$
$$\text{etc.}$$
$$\text{etc.}$$
$$\text{etc.}$$

Since a closed loop has no end, an indefinite number of rounds in this closed sensorimotor can be written as:

$$m_\infty = Op (Op (Op (Op (Op (Op (Op (Op \ldots .$$

Here m_∞ stands for both
  (1) the indefinite concatenation of operators, and
  (2) the resulting motor activity (behavior) of the organism after an indefinite application of operators on . . . what?
The question is both crucial and legitimate, for the primary operand m_0 (the primary motor activity constrained by "objections") has disappeared from this expression. This has significant implications for a theory of cognition and communication.

Until recently indefinite recursions were dismissed as dead ends indicative of misguided reasoning, and rightly so, for as long as solutions for such expressions were not forthcoming, their legitimacy should have been doubted. However, once the strategy for solving such expressions is understood, these difficulties disappear. This strategy is suggested in point 1 above: if m_∞ stands for the indefinite concatenation of operators on operators, then in the above expression the indefinite concatenation of operators on operators can be replaced by m_∞ at any point along this concatenation:

$$m_\infty = Op\,(m_\infty)$$
$$m_\infty = Op\,(Op\,(m_\infty))$$
$$m_\infty = Op\,(Op\,(Op\,(m_\infty)))$$
$$\text{etc.}$$
$$\text{etc.,}$$

that is, solutions for the above expression are all those values of m_∞ which, when operated on, produce themselves! Such a value is called an "eigenvalue" (the German *eigen* means "self"), which is consistent with the notion of a circular causality, for in order to close a loop, its beginning and end must meet or match up. An eigenvalue may also be seen as an *equilibrium* with respect to the operation Op, for if perturbed (pushed away from this equilibrium), the system will return to it.

Let me illustrate this with a simple example involving A and B. If Op stands for the operation of computing the square root, and if, for instance, in A: $m_0 = 95$, and in B: $m_0 = 0.03$, then the chain of events generated from these two initial conditions is:

	A			B	
m_1	$= \sqrt{95}$	$= 9.57$	$\sqrt{0.03}$	$= 0.17$	
m_2	$= \sqrt{9.57}$	$= 3.12$	$\sqrt{0.17}$	$= 0.42$	
m_3	$= \sqrt{3.12}$	$= 1.77$	$\sqrt{0.42}$	$= 0.65$	
m_4	$= \sqrt{1.77}$	$= 1.33$	$\sqrt{0.65}$	$= 0.80$	
m_5	$= \sqrt{1.33}$	$= 1.15$	$\sqrt{0.80}$	$= 0.90$	
m_6	$= \sqrt{1.15}$	$= 1.07$	$\sqrt{0.90}$	$= 0.95$	
m_7	$= \sqrt{1.07}$	$= 1.04$	$\sqrt{0.95}$	$= 0.97$	
m_8	$= \sqrt{1.04}$	$= 1.02$	$\sqrt{0.97}$	$= 0.99$	
m_9	$= \sqrt{1.02}$	$= 1.01$	$\sqrt{0.99}$	$= 0.993$	
m_{10}	$= \sqrt{1.01}$	$= 1.00$	$\sqrt{0.993}$	$= 1.00$	
m_{11}	$= \sqrt{1.00}$	$= 1.00$	$\sqrt{1.00}$	$= 1.00$	

$$m_\infty = \sqrt{1.00} = 1.00 \qquad\qquad \sqrt{1.00} = 1.00$$

Thus, independent of primary values, the values for A and B created by applying recursively the square root converge to the single equilibrial value 1, the eigenvalue of the "square root."

The eigenvalue "test" states that an operation applied to its eigenvalue must yield this eigenvalue. And indeed:

$$\sqrt{1} = 1.$$

This insight (or "solution") helps us understand the organism which recursively readjusts its behavior (operates on its motor activity) in accordance with limiting "objections" until a stable behavior is obtained. An observer watching this entire process, who has no access to the organism's sensations of the constraints on its movements, will report that the organism has learned to manipulate a particular *object* successfully. The organism itself may believe that it now *understands* (or has mastered the manipulation of) this object. However, since through its nervous activity, it has knowledge of its *behavior* only, strictly speaking these "objects" are tokens for the organism's various "eigenbehaviors." This suggests that objects are not primary entities, but subject-dependent skills which must be learned and hence may even be affected by the cultural context as well. A similar train of thought has been developed by Jean Piaget and his colleagues in Switzerland,[8] who have painstakingly studied the emergence of equilibrial behavior in infants which is indicative of an infant's grasp—note the reference to motor activity—of "object constancy."

What does all this imply for a theory of communication? Given the above, we can conclude that two subjects which interact with each other recursively will, *volens, nolens,* converge to stable eigenbehaviors which, from an observer's point of view, will appear as communicabilia (signs, symbols, words, etc.). Let me illustrate this with another mathematical example. Take the combined function of cosine and tangent, *costg*, as the recursive operator, and let the chain of events be initiated with an arbitrary value, say $m_0 = 0.5$:

$$
\begin{aligned}
m_1 &= \text{costg } (0.500) = 1.204 \\
m_2 &= \text{costg } (1.204) = 0.375 \\
m_3 &= \text{costg } (0.375) = 1.342 \\
m_4 &= \text{costg } (1.342) = 0.231 \\
m_5 &= \text{costg } (0.231) = 1.470 \\
m_6 &= \text{costg } (1.470) = 0.101 \\
m_7 &= \text{costg } (0.101) = 1.540 \\
m_8 &= \text{costg } (1.540) = 0.031
\end{aligned}
$$

[8]Jean Piaget, *L'Equilibration des structures cognitives* (Paris: Presses Universitaires de France, 1975), and B. Inhelder, R. Garcia and J. Vonèche, *Epistemologie génétique et équilibration* (Paris: Delachaux et Niestlé, 1976).

$$m_9 = \text{costg} \ (0.031) = 1.556$$
$$m_{10} = \text{costg} \ (1.556) = 0.015$$
$$m_{11} = \text{costg} \ (0.015) = 1.557$$
$$m_{12} = \text{costg} \ (1.557) = 0.014$$
$$m_{13} = \text{costg} \ (0.014) = 1.557$$
$$m_{14} = \text{costg} \ (1.557) = 0.014$$

.

.

.

$$m_{\infty_1} = \text{costg} \ (0.104) = 1.557$$
$$m_{\infty_2} = \text{costg} \ (1.557) = 0.014$$

A fascinating phenomenon—bi-stability—emerges: each stable value implies the other one:

$$\text{Op} \ (m_{\infty_1}) = m_{\infty_2}$$
$$\text{Op} \ (m_{\infty_2}) = m_{\infty_1}$$

Thus only when these operators operate in concert do we have a system with true eigenvalues:

$$m_{\infty_1} = \text{Op} \ (\text{Op} \ (m_{\infty_1})) = \text{Op}^2 \ (m_{\infty_1})$$
$$m_{\infty_2} = \text{Op} \ (\text{Op} \ (m_{\infty_2})) = \text{Op}^2 \ (m_{\infty_2})$$

If we take this as a metaphor for the interaction of two subjects, then the interaction becomes communicative if, and only if, each of the two sees himself through the eyes of the other. Note that in this perspective of communicative competence, concepts such as "agreement" and "consensus" do not appear and, moreover, need not appear (and this is as it should be, since in order for "consent" and "agreement" to be reached, communication must already prevail). These concepts, however, may very well appear in the vocabulary of an observer, who, outside the recursive loop watching the communicative interactions between the two subjects, sees no other way of explaining their concerted actions. But we should also note that, on the other hand, from this perspective, in which consciousness is attained only through conscience (that is, by identifying oneself with the other), communication, ethics, and love converge into the same domain.

Information and Communication in Poly-Epistemological Systems

MAGOROH MARUYAMA

Current theories of information and communication are marked by epistemological limitations, even fallacies. In this paper I will point out that since in practice communication takes place within poly-epistemological systems (epistemological structures vary from individual to individual, from group to group, and from culture to culture), the usual notion of what constitutes information is inadequate. A discussion of what is meant by "epistemologies" is best deferred until "information" and "communication" have been more clearly explained.

I. Classificational Information and Contextual Information

Elsewhere I have compared classificational information with relational and relevance information;[1] here I wish to compare it with contextual information.

Classificational information, the notion of "information" current in the U.S.A., has the following epistemologies or epistemological assumptions as its basis:

1. the universe consists of *substances* or *objects* which obey the law of *identity* and *mutual exclusiveness* and can be classified into a

---

[1] Magoroh Maruyama, "Communicational Epistemology," *British Journal for the Philosophy of Science*, 11 (1961), 319—27; 12 (1961), 52—62 and 117—31. Also see Magoroh Maruyama, "Metaorganization of Information," *Cybernetica*, 8 (1965), 224—36.

hierarchy of categories, subcategories, and supercategories;
2. the information value of a "message" increases with the categorical specification of the message; or a message's information value is greater if it describes an event which has a lower probability (for example, "it snowed in Florida in summer" has a higher information value than "it snowed in Quebec during the winter");
3. a piece of information has an *objective* meaning which is universally understandable, without reference to other pieces of information;
4. discrepancies within a message or differences between messages must have been caused by error; therefore, the discrepant portions should be discarded as inaccurate.

Contextual information, on the other hand, has the following epistemologies and epistemological assumptions as its basis:

1. the universe is basically heterogeneous;
2. the universe consists of interrelations and interactions of events, and everything occurs in a context which may vary from event to event; therefore, the value of a verbal or nonverbal message lies in its relation to its context, its interrelations with other messages, etc;
3. differences or disagreements within a message or between messages convey useful information because they reflect the richness of the interrelations, just as in binocular vision, the differentials between the two images enable the brain to compute an invisible dimension;
4. "objective" meaning is useless; there is no universal meaning; each piece of information must be interpreted in the context of other pieces of information and in terms of the given situation.

II. Non-informational Communication[2]

The transmission of information is not always the purpose of communication. In Danish culture, for example, the purpose of communication is frequently to perpetuate the familiar, rather than to introduce new information. A group of friends, for instance, may frequent the same coffee shop, eat the same pastry, and tell the same old gossip and jokes. Such preservation of the familiar is considered cosy (*hyggelig*).[3] New or

[2]Magoroh Maruyama, "Non-classificational Information and Non-informational Communication," *Dialectica*, 26 (1972), 51–9.
[3]Magoroh Maruyama, "The Multilateral Mutual Causal Relationships among Modes of Communication, Sociometric Pattern and Intellectual Orientation in the Danish Culture," *Phylon*, 22 (1961), 41–58.

different input may be considered disturbing. While in many other cultures persons express friendship by means of an exchange of opinions, feelings, and knowledge, in Denmark such modes of communication often tend to be regarded as aggressive, immature, impolite, and even hostile. It is considered injudicious to offer explanations or ask questions since the recipient may feel ignorant by implication; disagreement on factual matters is embarrassing because it implies that someone is wrong. (In this respect the Danes are almost the opposite of the Swedes.)

The Danes do not have the monopoly of non-informational communication, however. In the Chicano culture, conversation may simply follow the rhythm of manual work, with no information purpose.[4] In Denmark, a person who has no information may supply wrong information in response to a request in order simply to indicate his willingness to help.[5] In other cultures a disagreement may be an attention-seeking device rather than the expression of hostility which it appears to be.

III. The Non-Communicational Message

In Danish culture a refined insult is one of which the insulted person remains unaware. When this is achieved, the person doing the insulting derives satisfaction from being superior, whether or not there is a third person present who understands the insult.

Other intricacies of communication, such as criticality resonance and relevance should also be considered.[6]

IV. Criticality Resonance

In certain oppressive environments, dangers lurk of which the middle class is unaware, largely because it is unable or unwilling to acknowledge the reality of the oppression. This inability was an ingredient of intentional and unintentional racism. In the ghetto, for example, the legal authorities have often been illegal oppressors rather than legal protectors. Policemen and social workers have practiced favoritism in exchange for sex. Policemen have "confiscated" into their own pocket the ghetto residents' legally earned cash and possessions on the grounds of "suspicion" of theft. In such an environment, information such as a person's place of employment may be highly dangerous. Policemen may come to "arrest" a person for some suspected crime at the place where he

---

[4]Edward Hall, *Beyond Culture* (New York: Doubleday-Anchor, 1976).

[5]Maruyama, "Multilateral Mutual Causal Relationships."

[6]Magoroh Maruyama, "Epistemology of Social Science Research," *Dialectica*, 23 (1969), 229–80; and Maruyama, "Endogenous Research vs. Experts from Outside," *Futures*, 6 (1974), 389–94.

works. Even if he is later found not guilty, a few days of absence due to detention may have caused him to lose his job, because there were many other unemployed persons who could fill the job right away. More dramatic is the case of a woman from the ghetto who supplements her income from a low-paying job at City Hall by practicing prostitution. The fact that she is employed at City Hall could be dangerous information in the hands of her neighbors, who are unemployed and may covet her job. Should one of them learn of it, she might immediately inform City Hall of the woman's prostitution, thus causing the latter to be fired and perhaps obtaining the job for herself.

To a well-meaning person from outside who is unaware of this type of danger, such as a middle-class researcher or volunteer worker, ghetto residents must give some false information for self-protection, even though they may prefer not to lie.

The information giver's awareness of the information receiver's *un*awareness of the danger of information-giving is called "lack of criticality resonance." It causes unwilling distortion of information.

V. Relevance Resonance

If one is obliged to engage in a conversation whose purpose appears to be irrelevant, one will tend to supply distorted information. Prison inmates, for example, are frequently interviewed by psychologists, sociologists, students, or newspaper reporters whose purpose may be proving a theory, writing a book, getting a degree, or obtaining a promotion, increased prestige or income. These purposes are irrelevant to the inmates, whose chief concerns are better food, better and more up to date vocational training and less harassment from prison officials. The inmates, therefore, feel exploited by the interviewers who fail so obviously to address these issues, and in an effort to keep their intrusion to a minimum, they provide phony, though sophisticated, answers designed to keep the interviewers happy.

Each inmate has a well-prepared set of "answers" for each type of professional: one set for the psychologists, another for the sociologists, and so on. He can also create the illusion that the interview is highly relevant and that the interviewer is getting exclusive, "hot" information which no one else was able to obtain. Inmates can even fool psychiatrists. They learn the symptoms of mental illnesses and act them out before the psychiatrist, reducing them week by week or month by month in order to obtain an "improvement" report. Such counter-exploitation often occurs when there is no relevance resonance.

VI. Fallacy of Metacommunication Hierarchy

Still another factor affecting communication is the process described by Gregory Bateson as metacommunication.[7] This notion is commonly misunderstood as insinuation or indirect communication; Bateson, however, intended it to mean the comment on the truth mode of the communication. For example, if one were to write a letter, and then follow it with another declaring that the information in the first letter is false, the second letter would be a metacommunication on the first. A pathological situation occurs when a person becomes unable to tell whether he is lying to himself or not. A mother may wish to be rid of her child and therefore tell him that he is tired and ought to go to bed for his own good. The child may either acknowledge that his mother wants to be rid of him, which is painful, or in order to avoid this realization, he may convince himself that he is really tired and that his mother is being kind to him. If he adopts the latter course he is unwittingly lying to himself (and he should not realize that he is doing so). He develops an *ability to suppress* his own metacommunication to himself. If this occurs frequently, the child may develop an *inability to distinguish* his own metacommunication. The situation become pathological when the ability to suppress turns into the inability to distinguish his own truth mode.

Incidentally, Bateson's "double bind" is also often misunderstood. What he means by it is an *internal contradiction* within the *same* statement, *NOT* between two messages, as in the command: "Be self-assertive." It is *one* command which produces *self-contradiction* in the solution. If a person becomes self-assertive, he is obeying the command and is therefore not self-assertive.

So far, Bateson's theory is brilliant. The difficulty arises with his construction of a hierarchy of meta-levels: on communication there is metacommunication, on metacommunication there is meta-meta-communication, and so on. Contrary to Bateson's point of view, two communications can be meta-communication *on each other* in a *non-hierarchical* structure. For example, if George dislikes Tom and hits Tom, later telling him that it was only meant as a joke, then George's statement "It was meant to be a joke" is intended to be a metacommunication on his act of hitting Tom. However, the way George hit Tom leaves no doubt as to whether the blow was meant seriously. Therefore his act now becomes a sign that his statement is a lie.[8]

Consider another case in which the notion of hierarchy of meta-

---

[7]Gregory Bateson, Don D. Jackson, Jay Haley, John Weakland, "Toward a Theory of Schizophrenia," *Behavioral Science*, 1 (1956), 251—64.

[8]Magoroh Maruyama, "Basic Elements in Misunderstanding," *Dialectica*, 17 (1963), 78—92, 99—100.

communication is not useful. There was a man in China who had built a reputation for humility. Is his behavior on becoming famous a meta-communication that his statement "I am no good" is a lie? Or is his statement "I am no good" intended to be a metacommunication that his reputation is a false reputation? I think the falsehood, in either case, is merely a logical construct. In China, humility and the way to become humble are culturally institutionalized. Therefore no one was deceived either by his statement or by his reputation. Nor did he intend to deceive anybody. If anything, he was skillful at culturally institutionalized behavior, and he was admired for his ability to succeed in this pattern of behavior.[9]

VII. Epistemologies

Let us now take up the discussion of epistemologies which we had post-poned. I will not define here what is meant by epistemology because the concept of definition itself occurs only in one type of epistemology. Rather I will describe schematically four types of epistemologies, noting that there are many others as well as mixtures and overlappings (more thorough discussions of them can be found elsewhere[10]):

1. homogenistic, hierarchical, and classificational epistemology;
2. independent-event epistemology;
3. heterogenistic, reciprocally causal, homeostatic epistemology;
4. heterogenistic, reciprocally causal, morphogenetic epistemology.

The following tabulation can be regarded as an epistemological Ror-schach test: readers can identify their own epistemological assumptions by the way they react to the tabulation itself. Those whose thinking is dominated by a homogenistic, classificational epistemology will ask whether the categories in the table are universally valid, exhaustive, and mutually exclusive; finding that this is not the case, they will attempt to construct categories which are just that. Readers who assent to an independent-event epistemology will be uninterested in or even hostile to the tabulation. Finally, those who subscribe to homeostatic and morphogenetic epistemologies will consider the structure of the tabu-lation dependent on the context of given situations; they will regard the very interpretation of this tabulation as changeable and its meaning as flexible.

---

[9]Maruyama, "Basic Elements."
[10]Magoroh Maruyama, "The Second Cybernetics: Deviation-amplifying Mutual Causal Processes," *American Scientist*, 51 (1963), 164–79, 250–56; and Maruyama, "Paradigma-tology and Its Application to Cross-disciplinary, Cross-professional and Cross-cultural Communication," *Cybernetica*, 17 (1974), 136–56, 237–81.

Magoroh Maruyama

	HOMOGENISTIC	H E T E R O G E N I S T I C		
	Hierarchical classificational	Isolationistic independent-event	Reciprocally Causal homeostatic	morphogenetic
Philosophy	Universalism: Abstraction has higher reality than concrete things. Organismic: The parts are subordinated to the whole.	Nominalism: Only the individual elements are real. Society is merely an aggregate of individuals.	Equilibrium or cycle: Elements interact in such a way as to maintain a pattern of heterogeneous elements or they go in cycles.	Heterogenization, Symbiotization and Evolution: Symbiosis thanks to diversity. Generate new diversity and patterns of symbiosis.
Causality	Two things cannot cause each other. Cause-effect relations may be deterministic or probabilistic.	Independent events are most natural, each having its own probability. Non-random patterns and structures are improbable, and tend to decay.	Reciprocal: Counteracts deviation (negative feedback loops), both probabilistic and deterministic.	Reciprocal: Amplifies differentiation (positive feedback loops), both probabilistic and deterministic.
Logic	Deductive, axiomatic. Mutually exclusive categories. Permanence of substance and identity.	Each question has its answer unrelated to others.	Simultaneous understanding of mutual relations. No sequential priority. Definitions are mutual, not hierarchical. Logical values cannot be ordered.	
Perception	Rank-ordering, classifying and categorizing into neat scheme. Find regularity.	Isolating. Each is unique and unrelated to others.	Contextual: Look for meaning in context. Look for mutual balance, seek stability.	Contextual: Look for new interactions and new patterns. Things change and relations change. Therefore meanings change and new meanings arise.
Knowledge	Belief in existence of one truth. If people are informed, they will agree. There is the "best" way for all persons.	Why bother to learn beyond my own interest?	Poly-ocular: binocular vision enables us to see three-dimensionally, because the differential between two images enables the brain to compute the invisible dimension. Cross-subjective analysis	

Headings from Page 34 Continued throughout Table

Knowledge	Objectivity exists independent of perceiver. Quantitative measurement is basic to knowledge.			enables us to compute invisible dimensions. Diversity in perception enriches our understanding.
Information	The more specified, the more information. Past and future inferrable from present probabilistically or deterministically.	Information decays and gets lost. Blueprint must contain more information than finished product. Embryo must contain more information than adult.	Loss of information can be counteracted by means of redundancy or by means of feedback devices.	Complex patterns can be generated by means of simple rules of interaction. The amount of information needed to describe the generated pattern may be greater than the amount of information needed to describe the rules of interaction. Thus amount of information can increase.
Cosmology	Causal chains. Hierarchy of categories, supercategories and subcategories. "One-ness" with the universe. Processes are repeatable if conditions are the same.	The most probable state is random distribution of events with independent probability. Structures decay.	Equilibrium by means of mutual corrections, or cycles due to mutual balancing. Structures maintain.	Generate new patterns by means of mutual interaction. Structures grow. Heterogeneity, differentiation, symbiotization and further heterogenization increase.
Ethics	*Competition.* Zero-sum. If not homogeneous, then conflict. Let the "strongest" dominate homogenistically. Majority rule (domination by quantity).	*Isolationism.* Zero-sum or negative sum. Virtue of *self-sufficiency.* If poor, own fault. Do your own thing. Grow your own potatoes.	*Symbiosis: Static harmony.* Avoid disturbance. Restore previous harmony. Positive sum.	*Symbiotization: Evolving harmony.* Positive sum. Look for mutually beneficial relations with new elements and aliens. Regard differences as beneficial. Incorporate new endogenous and exogenous elements.

Religion	Monotheism: Creator and prime mover, omniscient, omnipotent, perfect god. Missionary work to convert others (belief in superiority of one's own religion).	Individual beliefs.	Static polytheism. Maintain established harmony of diverse elements.	Dynamic polytheism. Evolving harmony of diverse elements. Look forward to new harmony when new elements are added or new events occur.
Research rationale	Dissimilar results must have been caused by dissimilar conditions. Differences must be traced to corresponding differences in conditions which produced them.	Find out probability distribution and conditional probability.	Dissimilar conditions may lead to similar results due to asymptotic convergence to an equilibrium or cyclic oscillation. Find range of initial conditions which leads to same states. (*Note:* Equilibrium does not mean homogeneity. It means *maintenance* of heterogeneity.)	Similar conditions may produce dissimilar results due to differentiation-amplification. Find the amplifying network instead of looking for different initial conditions. (*Note:* Same amplifying network may produce different results. The difference is due to a disproportionally small initial kick which may be accidental. The amplifying network is more crucial than the initial kick itself.)
Objects	As volume and mass opposing empty space. Often mass represents human; space represents nature.	Each is its own expression.	Represents elements of universe in static harmony.	Represents growing process, vitality, vicissitude. Often the object is a condensation of spirits which permeate locality, and therefore represents space, not mass.
Time	Building is designed to persist in time. Permanence. Building is also considered to embody eternal principles.	Permanence is unimportant.	The form persists while the building materials may be renewed. *Example:* rebuilding of ISE Shrine every twenty years.	Increasing heterogeneity and symbiotization. Design permits addition of new elements, change in patterns, etc. *Example:* KATSURA Villa.
↓	Tension. Straight shaft	Caprice. Random.	Stability. Triangles of un-	Flow. Spiral or curve suggested by relative

Conceptual movement between objects	extending beyond object. Straight line spanned between points. Lines radiating from a point or converging to a point. Curves created by physical contour of buildings.		equal sides. Circular loops or oscillation keeps returning to the same points. *Example:* Some Japanese gardens.	positions of objects, but not by objects themselves. Curves extend and do not return. If returned, some change has occurred such as height. *Example:* some flower arrangements.
Perceiver's movement	Sequential. The perceiver maintains his identity as a point which moves in space.	Random.	Perceiver permeates simultaneously in all parts, and feels invisible parts and behind objects. The entire design is perceived simultaneously, not sequentially.	Generated by perceptual experience which has multi-branch effect. Not single-tracked.
Esthetic Principles	Unity by similarity, repetition and symmetry. Dominant theme is reflected in subdominant themes. *Examples:* Gothic. Islamic. French garden. Japanese Yamato culture. Some Japanese architecture shows hierarchical principles originating from Korea, homogenistic principles originating from China, and more recently from Europe and the USA.	Random, capricious, haphazard, self-sufficient and unrelated to others.	Static harmony of diverse elements. Balance. Completed equilibrium which cannot be disturbed. Design may be asymmetrical. Avoid repetition of similar elements. Perfectionist. *Example:* Japanese Yayoi culture.	Changing harmony of diverse elements. Avoid repetition and symmetry. Designed for simultaneously multiple as well as changing interpretations. Deliberate incompleteness to enable addition of new elements and changes. *Example:* Japanese Jomon culture.

The Japanese garden design and flower arrangement incorporate both homeostatic and morphogenetic principles in varying proportions, depending on whether the Yayoi principle or the Jomon principle is emphasized.

Space	Defined as between masses, between points, along shafts.	Unrelated units.	Miniature universe. Self-contained internal interaction of heterogeneous elements.	Locality with its characteristics (spirits or feelings) connected to other localities. Often thought of as around something (rock, grove) which is considered to be condensation of spirits permeating the locality.
House	Separate the inside from the outside.	Separate one household from another.	Continuation of outside into inside. Removable outer shells. Garden continues into house. River flows under floor. Floor extends to outdoors. Passive appreciation of nature and environment.	A base for activity and interaction with environment, such as Japanese farm house. Or can be a base for interaction with other households.
Room	Specialized rooms (bedroom, dining) occupied by specialized furniture (beds, etc.).	Specialized to individual user's needs rather than to generally categorized functions such as dining room.	Convertible. Furniture removable. Walls and partitions removable.	Convertible. Furniture removable. Walls and partitions removable.
Sociability	Socialize within homogeneous group. Hierarchy of groups, subgroups, and supergroups.	Self-containment. Freedom from social obligations and freedom for individual caprice.	Mutual dependency. Concern over not disturbing equilibrium. Perpetuation of familiar relations.	Make new contacts, new networks, change patterns of interaction. Go beyond own group.
Privacy	Own group's privacy against other groups. Group solidarity. Concern with patent, copyright, and other legal rights.	Individual insulation.	Maximum sharing of intimate concerns.	Less concern with privacy or property right, because interaction is seen as positive sum.

Activity	Hierarchically and homogenistically organized.	Up to the individual.	Nonhierarchical mutualistic activities to *maintain* harmony.	Generate new activities and new purposes through new contacts and new networks. Seek new symbiotic combinations and dissolve no-longer-symbiotic combinations.
Clarity	Neat categories, clear without context.	Individual meaning is what counts.	Contextually interpreted, interdependent meaning.	Ambiguity is basic to further development and change.
Design choice	There is "the best design" for all persons.	Individualized design.	Traditional designs are considered to be results of most satisfactory equilibrium.	Generate new designs by interaction in new contexts.
Planning	By "experts."	Everybody makes his own plan.	Plans are generated by members of the community and pooled together.	
Decision process	Majority rule consensus or "informing" the public in such a way that they will understand the "best" design.	Do your own thing. Let others alone.	Elimination of hardship on any single individual regardless of which way the decision is taken.	

VIII. Information, Communication, and Epistemologies

Psychological studies by O.J. Harvey indicate that about one third of white middle class Americans think and act in the homogenistic, hierarchical, and classificational epistemology, another third in other epistemologies, and the remaining third in various mixtures of epistemologies.[11] Rosalie Cohen has discovered that the percentage distribution of various epistemologies is different in other cultures.[12] Recent research

[11]O.J. Harvey, *Experience, Structure and Adaptability* (New York: Springer Publishing Company, 1966).
[12]Rosalie Cohen, "The Influence of Conceptual Rule-sets on Measure of Learning Ability," *Race and Intelligence, Anthropological Studies*, 8 (1971), 41−57.

findings indicate that there are culturally learned, physiologically measurable differences in brain functioning. Tadanobu Tsunoda[13] has found that natural sounds such as wind, waves, animal cries, bird songs, and insect songs have primarity in the dominant cerebral hemisphere in Japanese individuals, but go into the non-dominant hemisphere in Europeans. Japanese who were brought up in South and North America showed the same patterns as Europeans, and some (not all) Europeans who were brought up in Japan showed the same patterns as the Japanese.

The hierarchical epistemology tends to see humans as opposing nature, while in the homeostatic and morphogenetic epistemologies, humans are a part of nature and interact nonhierarchically with beings and forces in nature. Epistemological differences such as these may become physiologically measurable and identifiable in future research, adding to the evidence that epistemological differences exist.

IX. Conclusion

I have pointed out that the formulation and nature of "information" varies from epistemology to epistemology; hence poly-epistemological information systems would be more useful than the current mono-epistemological model. When communicating parties are unaware that they are using different epistemologies, they tend to consider each other to be unintelligent, illogical, deceptive, insincere, or unethical.[14] The seriousness of these consequences at both the personal and the political level makes the need for poly-epistemological information systems urgent.

---

[13]Tadanobu Tsunoda, *Nipponjin no Noo* (Tokyo: Taishuukan, 1978).
[14]Karl Mannheim, *Ideologie und Utopie* (Frankfurt am Main: Schulte-Bulmke, 1929), and Maruyama, "Communicational Epistemology."

"Voice" and "Print": Master Symbols in the History of Communication

ERIC LEED

W hen we reflect upon the nature of a coming postindustrial culture we invariably use terms that are thickly encrusted with symbolic significance. "Technology" is especially so. In cultural criticism, we use the term "technology" as a metaphor to deal with concerns central to Western civilization: how is individual autonomy gained? how is it lost? how is it to be regained? We use manufactured items to grasp ineffable and highly important issues: to what extent are human beings in industrial civilization gaining or losing control over themselves, nature, their destiny, the environment? Often the larger, metaphorical meanings of "technology" react upon these familiar things —television sets, cars, computers—and invest them with a deep moral significance. They become tokens of our liberties or enslavements.

Technology has such a privileged metaphorical status in cultural analysis because it has played an essential role in the West, not just as a weapon in the conquest of nature, but also as a means of social regulation. One of the distinctive features of European civilization is its increasing preference for technological and impersonal ways (as opposed to authoritative and personal ways) of establishing social discipline and rationality. Norbert Elias has shown how important the development of an "eating technology" was in the progress of European manners. Knives, forks, spoons, tableware, and the rules for the proper use of this equipment established the boundaries between individuals. Rules of

41

deportment, which in traditional societies might be prescriptive and proverbial, were in modern Europe absorbed into the techniques of using tools of consumption. Even before the modern period, technology—in the form of the technology of war—provided the ideal means by which the first autonomous individual, the medieval knight, won and defended his status as a free and independent person. Marx, for example, is in a long tradition when he theorizes about the relationship of industrial modes of production to the acquisition of a social self, personal dependence and autonomy. Although the cultural dominance of specific technologies might change, the role which technology plays in defining the boundaries between social selves remains the same.

Perhaps this explains why "technology" has such metaphorical potency: it became the preferred way of defining the social world, of governing the interpenetrations of enclosed, autonomous selves. Technology, especially when understood as a "machine," has been and continues to be the most prevalent way we reify mechanisms of self-regulation (the psyche, for example), which are the condition of our civility. In the seventeenth and eighteenth centuries the machine meant the "automaton," the self-regulating, self-running mechanism. It was the very image of the autonomous individual objectified and depersonalized. The automatic "man," like the machinery of the heavens, was a self-enclosed system which, if free from external interference, would maintain its regular and orderly motions. Here the image of the machine carried the positive values associated with a self-determining subject. But the definition of the machine changed in the nineteenth century. It was reconceived as a system of opposed parts so arranged as to transform raw energy into "work," a definition which coincided with a new conception of the psyche as a structure of entities in conflict, so arranged as to transform libidinal drives into "value"—art, science, civilization.

Once we recognize that traditionally the machine has been an objectification of the self-regulating psyche, we can also understand the ambivalence which industrialized man feels toward "technology." Through the metaphor of technology in general, we address questions about our inner state to the outer world. In our images of the machine, we project our attitudes toward those internalized structures of repression which both confine and focus our energies; we see our lives as the product of a peculiarly impersonal but human-created world which at once tyrannizes us and liberates our energies.

Unfortunately, very little attention has been devoted to the way in which "technology" has functioned in our discourse about the nature of industrial civilization and the direction it is taking. In this paper, I will argue that if in Western culture technology has long been a characteristic mode of establishing social distance, it is also the "distinctive feature"[1] of the way we think about the emergence of an industrial society. More specifically, I will explore the ways in which the metaphor "technology" has functioned in the history of communication and show how an opposition between individuality and community, symbolized by the terms "print" and "voice," emerges from it. Briefly, we should note that the symbolic significance of "print" and "orality" is often asserted to be (confused with?) an effect of the media: print, we read, individualizes its users, while oral/aural modes of communication communalize, acculturate, and integrate individuals into a tradition. Most histories of communication weave a narrative from these meanings which has all the explanatory power of myth: it charters contemporary institutions and shapes the moral stance we assume towards them.

The myth of communication tells the story of how defining elements have been introduced into our culture and suggests what is lost or gained with each technological acquisition. It can be told briefly. In the beginning was the spoken word, the performance within an immediate, face-to-face social context. Communication was a primary means of social and cultural integration. The nature of oral modes of communication was such that the tradition had to be continually actualized in the present by a performer. Music, in the comprehensive Greek meaning of the term, articulated a social and cultural space which enfolded both performer and audience. The work song could, for example, be regarded as the first "machine" of production, for it provided a way of integrating and harmonizing the labor of large numbers of individuals.[2]

But with writing, and more so with print, the performer was detached from his audience. Both the author and the reading public gained a certain distance from their community and from their tradition. The autonomy of individuals is often regarded as one of the effects of print. As Marshall McLuhan insists, "Print carries the individuating power of the phonetic alphabet much further than the manuscript culture could ever do. Print is the technology of individualism."[3] But this autonomy enjoyed by readers and writers could also be experienced as alienation. The distance from their social milieu of those who acquired their identity through literacy could encourage the hubris characteristic of the man of letters. He could believe that culture was something created by individuals—artists, philosophers, and critics—rather than collectivities.

---

[1] I am using "distinctive feature" in the sense defined by Jakobson and Halle: "Each of the distinctive features involves a choice between two terms of an opposition that displays a specific differential property, diverging from the properties of all other oppositions." Roman Jakobson and M. Halle, *Fundamentals of Language* (The Hague: Mouton & Co., 1956), p. 4.

[2] See Karl Beucher, *Arbeit und Rhythmus* (Leipzig: E. Reinecke, 1919), p. 399.

[3] Marshall McLuhan, *Gutenberg Galaxy: The Making of Typographic Man* (London and Toronto: Univ. of Toronto Press, 1962), p. 158.

He could believe that the culture of autonomous creators of texts was inherently superior to the culture of those who still relied on oral forms of communication. The medium of print could buttress the illusion that the interests of the men of culture and letters were, or should be, the interests and concerns of the culture as a whole. In the background of this narrative lurks Hegel's notion of the moral evolution of the individual and Marx's conception of revolution. Only here the "modes of communication" separate and differentiate individuals against the background of the collectivity, placing them in a position from which they can no longer recognize their social being.

This narrative provides us with the oppositions by which we understand the effects of the "mass media" upon contemporary societies. The present communications revolution is either the destruction of the individual "autonomy" caused by print, the closing of the critical distance necessary for rational political, economic, and aesthetic judgment. Or it is a new integration of divorced cultures, an end to the "alienation" which defined the sensibility of the man of letters, and a retribution against those who have presumed themselves to be above the totality.

The myth of communication does not, as I would like to show, come from thin air. It comes out of the dominant concerns of intellectuals engaged in understanding the processes which produced modern industrial civilization. But if the myth continues to function for us, it does so because it temporalizes the logic of our own cultural situation, articulating our present divisions of communicative labor as periods in time. Just as Freud analyzed the conflicts and identifications inherent in the bourgeois family, defining the logic of that situation, and deployed this logic in time as the history of civilization, we deploy those alternative forms of communication available to us—whether oral, literate, or electronic—as periods of history and a sequence of cultural norms. The myth of communication is an attempt to narrate the stratifications and lineaments of our own culture. In constructing this narration we have little choice but to use those symbols and motifs which the tradition supplies.

I. Oral Culture

The image of oral culture which dominates the works of McLuhan, Walter Ong, Eric Havelock, Alfred Lord, and others has a distinguished and complex history. It can be traced directly back to the general European reaction, in the late eighteenth and early nineteenth centuries, to the French Revolution and the ideas which were thought to have produced it. German and English thinkers like Fr. Schlegel, Fichte, and Edmund Burke sought to provide a non-revolutionary alternative to the French idea of the nation. They sought to root the nation state in solidarities which existed prior to any conscious acts of agreement, solidarities more binding than any of those contracts of association which liberals regarded as the basis of the social, economic, and political life. Those who rejected the liberal view of the society as artificial found their primary, pre-rational source of social cohesion in language, the encompassing linguistic milieu which defined the very identities of those operating within it.

The ideological reaction to the Revolution drew upon a new, revolutionary view of language initially put forward by Johann Gottfried Herder. In his *Über den Ursprung der Sprache* (1772), Herder advanced a seminal idea. Language was not, as Enlightenment thinkers had believed, merely a system of arbitrary signs useful for the transmission of ideas and the ordering of impressions. Rather, language was a primary mode through which a "self" is objectified and realized. Language is thus the precipitate of a human essence; the national tongue, the crystallization of a common identity. Charles Taylor has described this idea as the essential element in what he calls the "expressivist revolution," a revolution which utterly transformed thinking about society, history, personality, and art from 1770 to 1830. Taylor writes that language was not, to the men who effected this intellectual revolution,

> just a set of signs which have meaning in virtue of referring to something, it is the necessary vehicle of a certain form of consciousness, which is characteristically human. . . . In other terms, words do not just refer, they are also precipitates of an activity in which the human form of consciousness comes to be. So they not only describe a world, they also express a mode of consciousness . . . they realize it, and they make determinate what that mode is.[4]

Herder believed that the essential nature of language was best revealed in poetry and song and that the national language was not just a mode of description but a vehicle for the realization of the national self. In his essay on folksong, Herder insisted that oral poetry is inseparable from the "organic community." It was a particularly uncorrupted form of communal self-expression utterly unlike the educated poetry of the eye, cut off from music, from communal life and rhythms.

An entire generation of nationalistic, bourgeois intellectuals equipped with this conception of language, poetry, and the nation set about "discovering the people." Not surprisingly, the earliest collectors of folklore, men like K. Danilov in Russia, the Grimm brothers, and Brentano, described the folk as everything they were not—simple, illiterate, practical, and pre-rational. In the folk community any individuality was utterly absorbed into the collectivity. Out of this first discovery

---

[4]Charles Taylor, *Hegel* (Cambridge: Cambridge Univ. Press, 1977), p. 19.

of the people came the image of "oral culture" as a non-literate community which continues to resonate in folklore studies and anthropology.

This image of oral culture, the very antithesis of individualistic bourgeois society, has continued to stimulate scholars in our time. Milman Parry was not the only modern student of oral poetry who acknowledged his debt to the conception of the organic language community as continually actualizing itself in performance:

> The verses and themes of the traditional song form a web in which the thought of the singer is completely enmeshed. There is some strand of words to bind the lightest thought. . . . The old romantic notion of the poetry as a thing made by the people is by no means a completely false one. The poetry does stand beyond the single singer. He possesses it only at the instant of his song.[5]

In oral culture the poet cannot conceive of himself as an individual offering his own aesthetic creation. He is the person who makes present and real an always imminent tradition. In this context, "tradition" is less a constraining, suprapersonal authority than a set of tools, themes, motifs, and formulae which "enmesh" the thought of the performer and give his performance an immediate cultural validity; it is a structure of imminent speech acts coming down to the present from the decisions of past generations. This is a conception of tradition which would have been familiar to Edmund Burke:

> A nation is not an . . . individual, momentary aggregation; but it is an idea of continuity, which extends in time as well as in numbers and in space. And this is a choice not of one day, or one set of people or a tumultuary and giddy choice; it is a deliberate election of the ages and generations.[6]

The identification of tradition and community with spoken language has entered modern notions of oral culture through Saussure and Roman Jakobson. Jakobson writes, "the work of folklore is supra-personal and leads only a potential existence, it is only a complex of definite norms and impulses, a cosmos of actual traditions which the performers vitalize through the embellishments of their individual creativity."[7] Jakobson's description of the nature of creativity in oral cultures stresses the role which the collectivity plays as "censor" of the individual performer. What was too idiosyncratic to be integrated into the common stock of formulae

[5] *The Making of Homeric Verse: The Collected Papers of Milman Parry*, ed. Adam Parry (Oxford: Clarendon Press, 1971), p. 450.

[6] Quoted in Raymond Williams, *Culture and Society: 1780—1950* (New York: Harper and Row, 1966), p. 11.

[7] Roman Jakobson and Peter Bogatyrev, "Die Folklore als eine besondere Form des Schaffens," *Roman Jakobson, Selected Writings* (The Hague: Mouton & Co., 1966), IV, p. 6.

was forgotten; what was irrelevant or alien to the collectivity was not remembered. The writer of texts was inherently different from the oral poet. By virtue of the fixity granted his work by his medium, the writer could, if he chose to do so, create in opposition to his cultural setting in the hopes that some future generation might value his work. This choice was not available to the oral poet; nor was it possible for him to address an ideal audience or an audience separated from him in time. The "fixity" of oral song lay in the memory of the folk, the power of the singer lay in his ability to actualize that memory.

One can understand neither the original impulses which went into the construct "oral culture," nor the continuing relevance of this construct unless one remembers that it was fashioned by a literate intelligentsia, by men who had one foot in the republic of letters, the other in the "popular" tradition. In the oral folk they created their own counter-mentality, their own negation, and rooted this negation in the very medium of communication used by the people. Sound was regarded as redolent of affect, less distancing, rational, and superficial than vision. By virtue of the innately affective qualities of the medium, the oral poet could compell his audience to an unreflective identification with the elements of his song, and through his song, with the tradition which enmeshed his "slightest thought." This stylization of sound as a subrational medium fostering intuitive identification rather than judgment and reflection had a long career in the nineteenth century, particularly in musical aesthetics. It survives in psychology as the aural lobe which Freud affixed atop and to one side of his model of mind, a lobe with direct connections to the unconscious. These values invested in the medium reinforced the conviction that pre-modern mentalities were also pre-rational, pre-analytic, wedded to emotional immediacy and unfriendly to abstract thought. It is not unusual to find, in the context of communications research, the thesis that mentalities in oral communities are non-analytic *because* of the nature of the medium in use. Walter Ong argues, for example, that:

> There is no way for persons with no experience of writing to put their thoughts through the continuous sequence such as goes into an encyclopedia article. Lengthy verbal performances in oral culture are never analytic but formulaic. Until writing, most of the kinds of thoughts which we are used to thinking today would not be thought.[8]

Here sound, the voice, the very symbol of non-contractual communities, becomes the "cause" of the essential features of the mentality of those communities. Oral culture meant and continues to mean the thorough-

[8]Walter J. Ong, *Rhetoric, Romance and Technology* (Ithaca and London: Cornell Univ. Press, 1971), p. 2.

going integration of the cultural super-ego (the tradition), the individual
ego (the performer) and the emotional life of the audience. This image of
cultural integration was the ideal of men who experienced the massive
political, social, and economic changes of the nineteenth century and felt
themselves to be living in a disintegrating culture which was quickly
becoming a "past."

Dissolving under the impact of political and industrial revolutions,
"traditional society" was reborn as the construct of the "oral culture."
This construct was invested in the rural lower classes, the old regime, and
"primitive" non-European peoples. It is not surprising that folklorists,
sociologists, and anthropologists found in their research a thorough-
going confirmation of the ideals implicit in this construct of community.
Community existed before, below, and outside modern capitalist society.
Oral culture was a framing conceptualization of enormous usefulness to
men conscious of the modernity of their own age, for it pointed to the
pathologies inherent in industrializing societies while at the same time it
allowed social theorists to describe those pathologies as characteristic of a
transition period between two forms of community, one unconscious and
pre-industrial, the other consciously constructed and postindustrial.

The view of language and community implicit in "oral culture"
continues to promise fruitful ways of understanding the social function of
modern broadcasting media. James Carey, for example, has suggested
that the most appropriate vehicle for the understanding of "mass"
communications is a "ritual view" of the communications process:

> A ritual view of communication is not directed toward the extension of
> messages in space, but the maintenance of society in time . . . not the art of
> imparting information or influence, but the creation, representation and
> celebration of shared beliefs. If a transmission view of communication centers
> on the extension of messages across geography for the purposes of control, a
> ritual view centers on the sacred ceremony which draws persons together in
> fellowship and community.[9]

A ritual view of communication emphasizes precisely those functions of
language, poetry, and art which the earliest collectors of folklore located
in oral song. The image of oral culture, when applied to modern media,
highlights the belief that through those media we receive those essential
tools for the realization of social identities and relationships.

---

[9]James W. Carey, "Communications and Culture," *Communications Research*, 2, No. 2
(April 1973), 177.

II. Print Culture

In the history of communication, print is ordinarily conceived as one of the causes of modernization, or, more exactly, as a primary agent, condition, and vehicle of that "individualism" which is regarded as the psychic and social core of modernization. It is easy to find in the literature versions (though perhaps less blatant) of Walter Ong's thesis:

> The development of writing and print ultimately fostered the break-up of feudal societies and the rise of individualism. Writing and printing created the isolated thinker, the man with the book, and downgraded the network of personal loyalties which oral cultures favor as matrices of communication and as principles of social unity.[10]

None of this can, of course, be shown exactly. We have only to ask the question—"Why can't we say that print put the isolated thinker into contact with others, enmeshed him in a world of influences which transcended his locale, multiplying his memberships, allegiances and aversions?"—to recognize that we are in the presence of a thematic and ideological configuration which expresses our notions of what is involved in the modernization process. This process is most often described as one which creates "autonomous" individuals able to hold at a distance their community, tradition, and personalized forms of authority. In the history of communication "print" is a "summarizing symbol" which compounds and synthesizes a complex of ideas which we have used to describe the emergence of industrial societies.[11] I would like to sketch, in a general way, how these ideals mesh to form an explanatory structure. It will then be easier to evaluate those models for the transcendence of industrial society which we find in notions of a postindustrial culture.

One of the primary characteristics of print is that it is a neutral, impersonal, or "cool" medium of communication. Only thus can it serve as a channel through which private selves communicate without prejudicing their autonomy and independence. In effect, the impersonality of print is essential if communication is to serve as a "free" medium through which subjectivities negotiate. In historical terms this is best put by Elizabeth Eisenstein:

> Print is a singularly impersonal medium. . . . The publication in numerous editions of thoughts hitherto unthinkable involved a new form of social action at a distance. . . . The reading public was not necessarily vocal, nor did its members necessarily frequent cafes, clubs and salons of known political complexion. It was instead composed of silent and solitary individuals who were often unknown to each other.[12]

[10]Walter J. Ong, *The Presence of the Word* (New York: Simon and Schuster, 1970), p. 54.

[11]See Sherry B. Ortner, "On Key Symbols," *American Anthropologist*, 75 (1973), 1340.

[12]Elizabeth Eisenstein, "Some Conjectures about the Impact of Printing on Western

In ideological terms this idea was best stated by John Stuart Mill. The notion of "impersonal" and "neutral" media of communication meant something very specific to him, as it did to most Classical Liberals. It meant that the press, speech, and academic discourse should be free from domination by any specific cluster of social interests. Only then could these institutions function as a true "market" of opinion, becoming the structure through which individuals weave their private attitudes into a public opinion, a democratic version of the truth. Once "free," the media of communication would become like those other superstructural edifices—the law, state and religion—a means through which individuals harmonized their essential uniqueness.

These assumptions still provide the moral imperative with which we regard communication. We often insist that the media should be an impersonal means of social mediation. Criticism of modern broadcasting often comes down to the complaint that the media do not provide free access to private persons because they are dominated by corporate and bureaucratic elites who maintain themselves through the marketing of addictive symbols. The call for "alternate media" is inspired by concerns which are not so different from those of Mill: communications media should provide a means for the social realization of rational, autonomous individuals.

In order to understand the thematic configuration often symbolized by "print" but dominating our views of the function of communication in general, we must understand that complex of images, ideas, and attitudes which are represented by the term "individualism." Native to the nineteenth century, the term was first used by Joseph DeMaistre in the 1820s to describe the diseases of modernity—the exultation of the individual and the disintegration of social discipline. For DeMaistre and St. Simon, "individualism" carried purely negative connotations. It described those forces which had rent the dense networks of status and sociability of the Old Regime. Though he retained much of this negative sense of the term, Alexis de Tocqueville ascribed some positive moral value to the privacy of individuals and broke with the purely polemical meaning of the term. Tocqueville transformed individualism into an historical thesis which has proved astonishingly durable. For him individualism was both a cause and an effect of those social and psychological transformations which accompanied the transition from a hierarchical to a democratic society. It described what was left after the demolition of the ceremonial and legal supports of status which defined "aristocratic" society. This retreat of individuals into a world of sentimentalized family relationships created that vacuum between private roles and public powers which Habermas

Society and Thought: A Preliminary Report," *The Journal of Modern History*, 40 (1968), 53.

has called the "public sphere."[13] In describing the growth of this sphere in *The Fall of Man*, Richard Sennet draws upon Tocqueville's thesis. In *Democracy in America*, Tocqueville wrote,

> Individualism is a mature and calm feeling, which disposes each member of the community to draw apart with his family and friends; so that, after he has thus formed a little circle of his own, he willingly leaves society at large to itself.
>
> Aristocracy has made a chain of all the members of the community, from the peasant to the king; democracy breaks this chain and severs every link of it. . . . They owe nothing to any man, they expect nothing from any man; they acquire the habit of always considering themselves standing alone, and are apt to imagine that their whole destiny is in their own hands. Thus not only does democracy make every man forget his ancestors, but it hides his descendants and separates his contemporaries from him: it throws him back forever upon himself alone and threatens in the end to confine him entirely within the solitude of his own heart.[14]

Although modern historians of European culture might now quarrel with Tocqueville's terms, he has inclined us to interpret the essential processes of modernization as the growing individuation of opinion, the spread of anomie, the evacuation of the public sphere, and the pathos of private man.

The reality which "individualism" describes has been pushed back in time. C.B. Macpherson suggests that modern notions of the individual come from the idea of the "possession of the self," an idea native to seventeenth-century English political philosophy.[15] Some time ago Burkhardt found the image of the individual fully elaborated in the Renaissance prince and Ernst Cassirer located the roots of individualism in Renaissance philosophy.[16] We eventually "find" that individualism is older than the Renaissance, rooted in the Medieval concept of "lordship over the self," a concept which, in the early Middle Ages, was first appropriated by kings and the princely nobility, then by the middle nobility of castellans and barons, and finally by a youthful, landless, itinerant knighthood. Notions of the individual provide much of the intellectual continuity in European history. This history can be seen as the successive appropriation of this "noble" concept of the self-ruling

---

[13]Jürgen Habermas, *Stukturwandel der Öffentlichkeit* (Neuwied: Luchterhand, 1962).

[14]Alexis de Tocqueville, *Democracy in America* (New York: Vintage Books, 1961), pp. 193–94.

[15]C.B. MacPherson, *The Political Theory of Possessive Individualism* (Oxford: Clarendon Press, 1962).

[16]Jakob Burkhardt, *The Civilization of the Renaissance in Italy* (New York: Modern Library, 1954) and Ernst Cassirer, *The Individual and the Cosmos in Renaissance Philosophy* (Philadelphia: Univ. of Pennsylvania Press, 1972).

individual by inferior social orders, a process in which lordship over the self was believed to be gained through the exercise of the activity characteristic of the social group newly claiming this dignity. In the Middle Ages the status of the free independent individual was acquired and held through feats of arms, in the Renaissance through the practice of the arts, after the Protestant Reformation through work, barter, and the acquisition of property. Finally, in the nineteenth century, individuality was gained through the exercise of those political rights guaranteed to all citizens.

In *The Civilizing Process*, Norbert Elias provides a picture of the pyscho-social transformations underlying these revolutions of individualism.[17] The uniquely European civilizing process he describes is one in which external controls over social behavior were increasingly viewed as internal mechanisms of self-control. There is something uniquely Western and modern about the ways in which Europeans came to view social rationality as emerging from structures of inhibition, control, and restraint which were experienced as subjective and internal to the individual. Social rationality (and the rules which serve to subordinate short term pleasures to long term needs) was established in seventeenth and eighteenth-century Europe through a self-conscious reform of personality. For the lower orders this reform was accomplished by the establishment of prisons, workhouses, schools, standing armies, and permanent asylums for the insane. For the upper classes it was accomplished through books, sermons, and the general dissemination of models of civility. Only on the basis of "heightened emotional control, increased restraint of his emotions" could the individual view his acts, status, and identity as something minted out of himself.[18] This conception of the individual has remained a luminous ideal in our own time. Recently Isaiah Berlin defined individualism in terms of his own aspirations which are intrinsic to European tradition:

> I wish my life and decisions to depend on myself, not on external forces of whatever kind. I wish to be the instrument of my own, not any other men's acts of will. I wish to be . . . a doer-deciding, not being decided for, self-directed and not acted upon by external nature or by other men as if I were a thing, or an animal, or a slave incapable of . . . conceiving goals and policies of my own. This is at least part of what I mean when I say that it is my reason which distinguishes me as a human being from the rest of the world.[19]

This view of the individual, which went through successive stages of social dissemination in modern European history, required a new view of

---

[17]Norbert Elias, *Über den Prozess der Zivilisation* (Bern and Munich: Franke, 1969).
[18]Elias, "Introduction," p. ix.
[19]Quoted in Stephen Lukes, *Individualism* (Oxford: Blackwell, 1973), p. 55.

how identities were formed in society. No longer could the individual be seen as acquiring an identity through his formal relations to an external, objective moral order. This identity must be minted out of the self. Charles Taylor has suggested, too, that the modern notion of the self-defining subject required a new view of nature. The natural world was no longer a macrocosm of significances crucial to the correct conduct of human life. Now the world was a network of neutral and purely contingent relationships:

> . . . a disenchanted world is correlative to a self-defining subject. . . . The winning through a self-defining identity was accompanied by a sense of exhilaration and power. . . . The subject need no longer define his perfection or vice, his equilibrium or disharmony, in relation to an external order. With the forging of this modern subjectivity there comes a new notion of freedom, a newly central role attributed to freedom, which seems to have proved itself definitive and irreversible.[20]

If the modern, bourgeois notion of the individual required the desanctification of nature, it also required the desacralization of social relationships. The social identity of the individual was acquired through the ways in which he freely participated in the market of status. Correct social relationships had to be conceived as freely chosen by independent individuals establishing their obligations through a variety of social and economic contracts.

The emergence of a self-defining subject was early associated with literacy. By the sixteenth century, literacy had become one of the definitive signs—along with the possession of property and a permanent residence—of an independent social status. Literacy was symbolic not necessarily of wealth or class, but of the capacity to master a skill which entailed a degree of mental self-discipline. At the same time print was the medium through which new conceptions of the behavior appropriate for self-possessing individuals were disseminated. It was in printed form that countless manuals on manners, civility, dress, and deportment reached their audience. The hunger of the new media for content made the publication of exemplary materials from the desideratum of private lives profitable and popular. Print, Elizabeth Eisenstein suggests, provided the means by which a new view of the individual as author was distributed:

> . . . personal, private, idiosyncratic views of the author could be extended through time and space. Articulating new concepts of selfhood, wrestling with the problem of speaking privately for publication, new authors

---

[20]Taylor, pp. 8–9.

(beginning perhaps with Montaigne) would redefine individualism in terms of deviation from the norm and divergence from type.[21]

The role which reading played as a source of discipline, and the function of print as the vehicle for images of the individual, were appreciated by Jean-Jacques Rousseau. In *Emile* Rousseau preferred to spare his pupil the discipline of reading until the age of twelve, after the natural sources of Emile's autonomy were fully developed. The first book Emile was to read was Defoe's *Robinson Crusoe*, a work useful not because Emile was to live the life of solitude but because it clearly laid out the primary fiction by which Emile should judge all social relationships. He should assume the point of view of the autonomous individual and author of his world, the point of view of Robinson Crusoe, "in the evaluation of all other conditions. The surest way for him to rise above prejudices and to bring us his own judgements into line with the true relations of things is to put himself at the point of view of the solitary man and to judge everything as this man would with reference to reality."[22]

Rousseau also explains best the modern preference for impersonal and neutral structures which mediate relations between individuals, structures like the open market, nature, the state, the media of communication. The best guarantee of freedom and individuality was for all members of a society to be equally dependent upon a will which was itself indifferent to the artificial distinctions of status which men erect between themselves. To be dependent upon a personal will was slavery. But to be dependent upon a source of authority which was impersonal and neutral—whether the will of nature or the artificially reconstituted "nature" of the General Will—was to enjoy the utmost degree of liberty attainable in civil society.

When the basic social unit came to be regarded as the independent individual, social relationships and—more important for us—communicative relations had to be judged in new ethical terms. The domain of letters was the domain of freedom, that impersonal and empty space, that neutral matrix in which individuals formed their own judgments, independent of those institutions which enforced political and religious orthodoxies. Impersonal modes of communication were essential if the individual were to retain and develop his autonomy through the chosen means of expression. The new view of the self-determining subject provides the ideological and moral imperative behind what James Carey calls the "transmission view of communication." This view, which dominates communications research, especially in the United States,

[21]Eisenstein, p. 23.

[22]J.-J. Rousseau, *Emile*, ed. and trans. W. Boyd (New York: Columbia Univ. Press, 1963), p. 84.

centers on the belief that communication consists of channels for the transmission of information across space. This process of transmission is evaluated morally in terms of how it develops the conditions for independent judgment or how it qualifies the autonomy of individuals becoming an instrument of "social control." This view pervades our thinking about the communications process regardless of one's political stance, for it rests upon the fiction of the autonomous individual, the fiction which provides the continuity of modern Western culture.

III. The Fixity of Print and the Literate Identity

Most of the effects of revolutions in the history of communications can be traced to the "fixity" which printing and print lend to discourse. Writing makes the word durable. The power of the printing press to duplicate texts infinitely enhances the chances that any given text will survive. This fixity has implications for the ways in which literate men regard their culture and their own identities. For the fixity of writing and print permits the amassing of a cultural tradition and gives the tradition a permanence and an apparent autonomy of its own. We recognize that the adoption of the alphabetic script by the Greeks did not "create" a new, literate society. Paradoxically the use of a more efficient writing system by the Greeks made it possible to fix and consciously address the ways of thought and speech identified with oral discourse. As Eric Havelock writes, "the alphabet's intrusion at this point into the history of Homo Sapiens introduced not literacy but a permanently ingrained and complete record of the ways of non-literacy."[23] A similar thing occurred, Elizabeth Eisenstein argues, in the "print revolution." Print did not so much create a "print culture" as provide the means for the preservation and consolidation of a manuscript culture. The effects of both writing and print upon the mentalities of those who used them were indirect but nonetheless pervasive. With the consolidation of the oral and written traditions, these traditions no longer totally "enmeshed" the thought of those charged with their preservation and explication. The received tradition could be perceived as objective fact, its inconsistencies and correlations noted and examined. Moreover, with the consolidation of the "old" tradition in print, it was possible to determine what was "new" and to see the history of art and science as a "progressive" apprehension of previously unsuspected truths.

The durability of the printed word seemed to give the written tradition an independence from any particular social or cultural locale, any discrete age or place. Paul Ricoeur suggests that writing lends

[23] Eric Havelock, *Prologue to Greek Literacy* (Cincinnati: Univ. of Cincinnati Press, 1971), p. 9.

discourse a peculiar autonomy, and it is by analogy with the "autonomy of texts" that intellectuals begin to talk of the "autonomy" of super-structures and cultural traditions in general. In the mass of inscriptions a world of meaning is created which is parallel to, but nowhere wholly coincident with social and cultural realities. The permanence and autonomy which writing and print lend to the process of signification even severs the meaning of a given text from the intentions and motives of its author. A text may contain coherencies, contradictions, assumptions which are wholly unintended by its author. Ricoeur explains, "with written discourse, the author's intention and the meaning of the text cease to coincide. This dissociation of the verbal meaning of the text from the mental intention of the author is what is really at stake in the inscription of discourse."[24] The divorce of the written text from its author becomes even more radical when one begins to talk of the entire corpus of work done by a given author and the relations of this corpus to the work of other authors. The coherence of the written tradition cannot be seen as something "intended" by those who have contributed to it. The written tradition organized and broadcast through print was an autonomous world which enfolds writers and readers in complex webs of meaning spun over a span of time measured in centuries and millenia. Because of the autonomy of texts and the much more obvious autonomy of the entire tradition from any particular social context, writing and print were seen as media of transcendence, a means by which individuals could remove themselves in mind from their social place.

This severance of superstructures, of the world dependent upon documents, duplicable texts, and visual representations, from infrastructures, from the activities in which men reproduce the material conditions of their lives, is generally thought to be one of the characteristic marks of modern societies. It is in contrast with this autonomy of superstructures that historians of communications emphasize the total integration of super- and infrastructures within oral cultures. There, one can find no divorce between meaning, act, tradition, and audience. Walter Ong insists that the meaning of the voiced word depends upon the resonances emanating from the context in which it is spoken: "The word as record depends for its meaning upon the continuous recurrence of the word as event."[25] Every word, gesture, and image has meaning only within the circle of auditors.

The autonomy and permanence of the written tradition have implications for the identities of those who realize themselves through the world of texts. The written tradition provides a highly ramified set of

[24]Paul Ricoeur, "The Model of the Text: Meaningful Action Considered as Text," *New Literary History*, 5, No. 1 (1973), 103.
[25]Ong, *Presence of the Word*, p. 33.

trails into the past, a network of frozen significations which can "remove" the man of letters from his social place, his real world. Perhaps the best evocation of the pathos inherent in the psychic situation of those who lose themselves in the domain of texts is to be found in Milman Parry's reflections upon the character of those who seek an alternative "past" through the labyrinths of the written word. Parry recognized that his interest in oral cultures stemmed from the longing for a communicative experience which was utterly different from the self-estrangement characteristic of the wholly literate man:

> When one trained in [the historical method], while still staying in the past, turns his eyes back to his own time, he cannot prevent a certain feeling of fear—not for the fact that he has become a ghost in the past—but because of what he sees in the person of his living self. For in the past where his ghostly spirit is, he finds that men do the opposite of what he has been doing: they, by their literature, make it a mirror for themselves, and as a result, the past has a hold upon them which shows up the flimsiness of the hold which our past literature has upon ourselves.[26]

A peculiarly indefinite status is conferred upon the scholar or the intellectual by his membership in the autonomous domain of the inscribed word. His liberation can also be experienced as dismembership, his freedom as a divorce from any determinative cultural tradition. Here we find the image of the man of letters as an individual removed from his "life," a "ghost in the past," a person who occupies a liminal space somewhere between his social and cultural origins and the object of his intellectual journey. This sense of dismembership or psychic divorce from any concrete time or place (the characteristic self-consciousness of the man of letters) is the context of his longings for membership, integration, and absorption in a morally pure community. These qualities of "wholeness" are characteristic of "oral culture" and personified in the figure of the "oral poet." Oral culture was the curative of those who experienced themselves as liberated from their time and place, suspended in the world of inscribed discourse, the idealized childhood of the literate man, his longed-for past. But this past was also useful in describing a future community in which his status-lessness, his rootlessness was to be resolved. In the images of the future which we find embedded in notions of postindustrial culture, the contradictions which defined the consciousness of men in modernizing society seek some kind of resolution: the dilemmas of membership and autonomy narrated in the myth of communication are reconciled in the visions of transcendence

[26]Milman Parry, "Historical Method and Literary Criticism," *The Making of Homeric Verse*, p. 410.

dreamt by men who felt that they lived in the belly of an industrial leviathan.

IV. Postindustrial Culture

I have argued that "print" and "voice" are summarizing symbols for the opposition between individuality and community which pervades our notions of the nature of the modernization process. It is difficult for natives of the modern West to assume a place outside the thematic and ideological configuration repesented by these symbols, for it supplies us with our notions of transcendence. The future is the stage upon which we arrange those figures that have played successful roles in the moral career of Western man. There Lewis Mumford places "Damnation," Buckminster Fuller, "Salvation." In the technological society, the lineaments of which we are beginning to discern, Jacques Ellul sees the coming of the Antichrist, Marshall McLuhan the restoration of the truly human. We must not take these futures too seriously but neither must we dismiss them. In the future, that not-yet-specified dimension of time we may recreate with the operational motifs of our culture, and by manipulating them, come to know them better.

Most visions of postindustrial culture stress the centrality of communication and suggest that the media of communication will play a much greater socially and culturally integrating role than they have in the past. Victor Ferkiss argues, for example, that the "single factor that most distinguishes the coming civilization is the substitution of 'communication' processes for traditional 'work' as man's primary activity."[27] At the same time communication will function less as a means for the production, preservation, and transmission of information and more as a means for "affecting states of consciousness" in the direction of more intense social and cultural interaction. McLuhan and Ong interpret the present and future through the lens of "oral culture," but they find essentially the same thing. They are convinced that we are in a period of "secondary orality" or "literate orality" in which the electronic broadcasting media will shape mentalities in much the same way as the oral media of communication shaped the mentalities of the pre-literate folk.

Even critics of the media, those inclined to see the media as just one more of those structures of repressive desublimation, agree with this thesis, though they find the coming age of electronically enforced community distasteful and inimical to that ideally rational, critical public associated with the age of literacy. George Gerbner argues, for example, that television most spectacularly makes the mental state of modern

[27]Victor Ferkiss, *Technological Man: The Myth and the Reality* (London: Heinemann, 1969), p. 108.

audiences more and more like that of archaic publics: "The nearly universal, non-selective and habitual use of television fits the ritualistic pattern of its programming. You watch television as you might attend a church service, except that most people watch television more religiously."[28] It is clear that if there is a consensus about the nature of a postindustrial culture it lies in the notion of the centrality of the communications processes and the function of these processes in establishing cultural cohesion.

But how are we to interpret the consensus? It is very difficult to measure "degrees" of socialization and acculturation, even more difficult to demonstrate how television actually affects the emotional states of audiences, and almost impossible to determine the relative degrees of something like social integration on an historical scale. No culture can survive for more than a generation unless it has effective mechanisms for the production and dissemination of the symbols which define the sources of solidarity and the differences of age, class, sex, and dignity. How binding and determinative these symbols are, how symbolic representations of identity actually operate in behavior—these are the questions which most historians, anthropologists, sociologists, and literary critics prefer to "put aside" for the moment. Unquestionably what we find in the modern world is the proliferation of mechanisms for the dissemination of the symbols which give social meaning to our memberships and individualities. But this proliferation may mean that cultural superstructures weigh more heavily upon our private selves, or it may mean that the profusion of choices, the profusion of "inputs," preserves our distance, our freedom, our autonomy from the sources of communication.

There is also another issue, in a more historical vein. Are we to see the centrality of socially integrating media in our society as a signal for the closure of an "industrial era," or as the cultural fruition of a protracted, many-phased industrial revolution? Raymond Williams believes, and I with him, that in the present revolution of communications we can see a process in which an industrial "society" is acquiring all the characteristics of a "culture." And this is not an accidental process. The dominant concern of nineteenth-century social theorists—whether radicals or reactionaries—was with the sources of community. These concerns were shared by engineers and technicians and by entrepreneurs who saw a potential market in the obvious need for more effective cultural ties, more intimate human bonds. a more genuine solidarity than seemed available in industrializing societies. The awareness of the need for new mech-

---

[28]George Gerbner and Larry Gross, "Living with Television: The Violence Profile," *Journal of Communications*, 26, No. 2 (1976), 177.

anisms through which people could be put in touch with each other, entertained, and taught was a significant factor in the development of our present technologies of communication. Williams writes,

> In many different countries, and in apparently unconnected ways, such needs were at once isolated and technically defined. . . . It is especially a characteristic of these communications systems that all were foreseen—not in utopian but in technical ways—before discovered and refined. In no way is this a history of communications systems creating a new society or social conditions. The decisive and earlier transformations of industrial production, and its new social forms, which had grown out of a long history of capital accumulation and working technical improvements, created new possibilities, and the communications systems, down to television, were their intrinsic outcome.[29]

The content of these new media—film, radio, and television—was not drawn from the culture of elites but from the culture of the "masses" (whom Williams defines as "people we don't know yet"). Sports, popular melodrama, dance, game shows, music hall variety, and the entire range of popular music from blues, country, gospel and their various permutations, to big-band music, filled the "air-waves" and continue to do so. Though this content is often despised as simple, dumb, and naive—and much of it is—by traditional, literate elites, it "works" and is popular, and does not apparently require anything more than purely commercial support.

Most often the present revolution in communications is understood as having integrating or "massifying" effects. Almost no one whom I know claims that the "mass media' will create new forms of autonomy, individuals who are adept at crafting the materials of popular culture, turning them into "art." But this possibility must be taken seriously, if only because of what we know about communications revolutions in the past. The introduction of writing and print created the conditions in which those responsible for the maintenance of cultural traditions could organize, amass, analyze, and ultimately manipulate those traditions, creating new forms of representation and thought. With writing, the Greeks, and with print, the Humanists were able to become conscious of the logic and the illogicalities of the inherited culture and to draw up the rules for thinking, speaking, sculpting, building, and healing.

We may find that the present revolution in the means of communication will have a similar effect. Through video-cassettes, tapes, records, film, we can collect a musical, dramatic, and visual heritage which assumed a definite shape in the first half of the twentieth century but

[29]Raymond Williams, *Television: Technology and Cultural Form* (New York: Schocken Books, 1974), p. 19.

whose roots reach deeply back into the popular styles of European and African art. It is clear that, at present, performers like Ry Cooder, Leon Redbone, and the Pointer Sisters are "collecting" the musical styles which emanated from industrializing America and are playing these styles with a fidelity and grace which few of the "originals" could attain. Past communications revolutions have often presaged "classical" periods of cultural development, periods of intense creativity. These periods were predicated upon the existence of new means for the consolidation and organization of the "old." New media of communication always created the conditions in which men could address their culture as conscious, rational individuals engaged in the recombination, the reintegration of its elements. It would not be surprising if individuals in a postindustrial culture were to engage in this kind of activity with respect to that often rich, meaningful, and emotionally powerful tradition of popular culture which has been so widely disseminated in the last three decades.

II
Technology,
Mass Culture, and the
Public Sphere

Mass Media: Tools of Domination or Instruments of Liberation? Aspects of the Frankfurt School's Communications Analysis*

OSKAR NEGT

If I believed that our present bourgeoisie were going to live another hundred years, then I would be certain that it would continue to babble on for hundreds of years about the tremendous "possibilities" that the radio, for example, contains. . . . I really wish that this bourgeoisie would invent something else in addition to the radio—an invention that would make it possible to preserve everything that the radio is capable of communicating for all time. Future generations would then have the opportunity to be astounded by the way a caste made it possible to say what it had to say to the entire planet earth and at the same time enabled the planet earth to see that it had nothing to say. A man who has something to say and finds no listeners is bad off. Even worse off are listeners who can't find anyone with something to say to them.

Brecht, *Radio Theory* (1927)

*For a full exposition of the relationship between critical theory, the public sphere, and culture, see Peter Uwe Hohendahl, "Critical Theory, Public Sphere and Culture. Jürgen

65

I

The articles published on mass media in the *Zeitschrift für Sozialfor-schung* during the 1930s are thematically heterogeneous and operate at various levels of argumentation. If we were to gather works dealing with communications theory as a collection of essays representative of the Frankfurt School, then two things would have to be made clear from the start.[1] First, we are dealing here with bulky goods, with opposite positions in a concrete historical situation—positions that can be integrated into the current appropriation process of leftist literature and into the existing academic apparatus only by totally falsifying their contents. Secondly, these essays are informed by a pervasive cognitive motif which consists of the critique of culture, communication, and mass media as measured against the historical level of fulfillment of objective human needs. Thus, a suitable appropriation of these essays is possible only from a *political* position which takes a stand for people's needs, interests, and strivings towards autonomy and which also consciously undertakes practical steps towards making these things a reality today.

This necessitates the setting of limitations, not as external delinea-tions of position, but rather to designate the comprehensive framework in which a meaningful treatment of problems is possible and necessary. That is, a treatment which takes genuine social significance as its gauge. If we were to believe the literary and scientific activity of recent years that has been oriented towards concepts of language and communication, where language has assumed growing importance in solving social and individual biographical conflicts. The opposite is true. To be sure, research projects on language barriers, the estalishment of professorships in sociolinguistics and communications research, the analytical philos-ophy of language, and even inventiveness in the development of models for speech therapy, have been encouraged. On the other hand, we have the actual social condition of communication's speechlessness; the potential for violence, and the manipulative use of language as a tool of

Habermas and his Critics," *New German Critique*, 16 (1979), 89−118. An earlier version of this paper was given at the International Symposium on Postindustrial Culture: Technology and the Public Sphere, held at the Center for Twentieth Century Studies, The University of Wisconsin-Milwaukee, November 1977.—*Ed.*

[1] These essays have indeed been collected in a volume entitled *Kritische Kommunikations-forschung*, ed. Dieter Prokop (Munich: Hanser Verlag, 1973). My present essay appeared as the introduction to this volume under the title "Massenmedien: Herrschaftsmittel oder Instrumente der Befreiung?" and in translation in *New German Critique*, 14 (1978), 61−80. A systematic overview of the authors and most important works of the *Zeitschrift für Sozial-forschung* that also clearly delineates the theoretical development may be found in Alfred Schmidt's introduction to the reprint of the *Zeitschrift für Sozialforschung* (Munich: Koesel Verlag, 1970).

domination. The contradiction between these two sets of phenomena has never been as great as it is today. The daily experience of this contradiction's insolubility under the ongoing conditions of a bourgeois public sphere with a fixed framework leads critical reflection about communication and language to assume the character of a supplementary ideology inasmuch as it is generally separated from political contexts. This is the idealizing complement of a society whose guiding institutions are geared precisely to hinder communication and verbal articulation.

The totalizing tendency of language and communication may have to do with their very structure; yet, as *elements* they are undoubtedly constitutive of historical praxis. As a rule, however, communication will take the form of cyclical conversations if verbal understanding dissociates itself from the context of social praxis. This is true for verbal understanding in terms of epistemological comments on hermeneutic rules of explication and competence of language, as well as in terms of everyday consciousness, for example in conflict situations in the family. Since in its own terms, however, every conversation appeals to autonomy, it is an indication of a false, twisted consciousness if it does not at the same time name its own particularity: that which is nonverbal, the mute coercion by the basic circumstances of violence and domination. And not only that. Emancipatory communication also necessarily designates the objective conditions under which the human being can become more of a subject and can build more autonomous and more comprehensive relationships to reality.

Before we concern ourselves with mass media and verbal communication, let us address the concept of communications itself. What is almost forgotten today was formulated with uncompromising precision by Max Horkheimer over thirty years ago: not only is the content of communication and the media operating under capitalist conditions of utilization subject to criticism, but rather it is the intrinsic dialectic of the concept of communication that must be unfolded as a category of reality. Like all other historical substantive concepts, this one, too, has the tendency to turn into its opposite under certain social conditions. Kant, in guaranteeing his aesthetic power of judgment a degree of intersubjectivity and objective liability that would correspond to the laws of natural science, could still rely on a colloquial *sensus communis aestheticus*. Dewey, over a hundred years later, was able to celebrate art as the most universal and freest form of communication. In contrast, Horkheimer verified the objective collapse of all connections between *experience* and *communication*. For Horkheimer, this split meant among other things that it was impossible to mediate authentic possibilities for expression of experience via mass communication: "Europe has reached a stage in which all highly developed means of communication serve to fortify the barriers 'which separate human beings from each other.' The radio and movie theater are just as effective here as airplanes and cannons; the way people are today, they understand each other only too well. If they ever started not understanding each other or themselves any more, if they become suspicious of their forms of communication and the natural seemed unnatural to them, then at least the horrifying dynamics would come to a stop. Whatever is still communication in most recent works of art only denounces the prevailing forms of communication as the instruments of destruction and harmony as a mirage of decay."[2]

This statement by Horkheimer shows how crucial a concept of the concrete totality of society is: a dialectical social theory must be conscious of the decisive historical patterns of development and combine the description of the empirical condition of human beings with studies of objective class interests. It is impossible to do without such a theory if we want to prevent cognition from being reduced to the observation of superficial, abstract, isolated appearances.

Just as the essence of communication described by Horkheimer is revealed as absence of communication, the essence of the relationship among today's mass media becomes discernible if one looks at the state of literature, the educational systems, the level of cultural development, and the current state of class struggles, rather than at a media researcher's specialized topics of investigation. This is a fundamental theoretical question, not a question of cleverly constructing hypotheses nor of the individual researcher's personal capability. This is precisely what Paul Lazarsfeld meant when he summarized the difference between "administrative and critical communications research" with the provocative comment that the latter concerned itself essentially with those ideas, initiatives, and modes of behavior that force their way into the media in totally distorted form if at all.[3] *The media do not constitute the core of a critical media theory.*

In contrast to this, a branch has recently been established in academia in which theory at best is accepted in the guise of Merton's theories of medium range or of structural-functional analysis. In effect, however, empirical, individual research prevails in this branch. Dieter Prokop has classified this orientation as positivistic media research.[4] For the limited

---

[2]Max Horkheimer, "Neue Kunst und Massenkultur," *Kritische Theorie*, II (Frankfurt am Main: Suhrkamp Verlag, 1968), p. 320.

[3]See Paul F. Lazarsfeld, "Bemerkungen über administrative und kritische Kommunikationsforschung" in *Kritische Kommunicationsforschung*, pp. 7–27.

[4]Dieter Prokop has compiled the most important works in this line of research and criticized them in extensive, thorough introductions in two anthologies. See *Massenkommunikationsforschung 1: Produktion*, and *Massenkommunikationsforschung 2: Konsumtion* (Frankfurt am Main: Fischer Verlag, 1972, 1973).

purpose of this essay it is not necessary to present and criticize its methods and research results in detail. It may therefore seem abstract and extremely simplified if we confine our criticism here to the famous Lasswell formula, which appears not coincidentally, more or less modified, in almost all positivistic mass communications research studies. With incomparable precision it leads to the concentration of divisions of scientific labor. For example, it narrows the field to be studied in such a way that one no longer recognizes the preliminary abstractions which have been applied to social totality for efficiency's sake, due to the formal consistency of the final study. Lasswell's formula reads: "who says what in which channel to whom with what effect?"

It is clear how this formula makes it easier for the media researcher to gain an overview of the object under investigation. It enables the researcher to proceed according to the standard rules of specialized topics of study and modes of questioning, all within an academia cut off from the overall social process of production and reproduction. Despite any methodological and objective differentiation that may convey the impression of a particularly rewarding increase in knowledge via exact measurements of ratios and via correlations of a variety of variables, such as transmitter, receiver, preferences, attitudes, behavioral patterns, etc., there are two assumptions that remain constant throughout the individual segments of disjointed communications research (message-, carrier-, receiver-, effect research, etc.). These assumptions are first, that the object under investigation, in this case the media apparatus as well as receiver, is left unchanged by "value free" research. Second, that the researcher's professional role is a static one: the researcher considers him/herself exempt from the conscious decision for social interests and historical tendencies because the methods guarantee the objectivity of knowledge for the researcher alone.

Both assumptions are indicative of deception conditioned by interest. Scientific investigation that foregoes the opportunity to take a determined and conscious stand for the better possibility inherent in every object of cognition over bad reality does *not* in *any* way leave (untouched) the object which it has raised to the level of communicable knowledge. Rather, this renunciation serves merely to facilitate the extension of existing patterns of domination through new possibilities for utilization. The researcher faces this alternative with every new investigative step, and if the researcher invokes the objectivity of knowledge guaranteed solely by methods, he/she has already decided for a certain class interest. *Positivistic communications research is essentially utilization research.* By virtue of its inability either to comprehend or practically to solve the contradiction between technical possibilities and outdated systems of production that is reproduced in the media in a way specific to them, positivistic communications research becomes the administrator of remnants of a bourgeois public sphere that finds itself in an irretrievable process of decay. This bourgeois public sphere has, in addition, turned the *citoyen*, on whom it once relied, into a consumer, who sees the path to the television knob as the way to freedom and autonomy.

II

It was not intensified theoretical study, but rather, a political rebellion in West Germany which destroyed the illusion of a liberal public sphere that provided all social interests with an equal voice. In the beginning this rebellion of the late 1960s made use of the media to create its own public sphere by countering the secretive practice of university administrators and the concealment of the total oppression of third-world peoples. Such a demonstrative public sphere was intended to draw the masses into a process of enlightenment which would in turn convince them of the obsolescence of existing power relationships. Confronted with the joint efforts of subliminal coercion by mass media and manifest coercion by police—i.e., personally experienced one-dimensionality of the system— the West German protest movement turned to works by Wilhelm Reich, Horkheimer and Adorno, and Herbert Marcuse to legitimate its praxis. This was of fundamental significance for the beginning phase of the movement. The anti-institutional and anti-authoritarian element in the revival of critical theory fused with the attempt, via politicization of interests and needs, to accomplish three things: 1) to break through the compulsive and pervasive mediations of commodity exchange; 2) to break through the violence latent in the mechanisms of instrumental reason and structurally inherent in the sublimation and repression of basic instincts; 3) to establish meaningful immediacy, in which the split between communication and experience is in turn eliminated. Pirate editions of works by Reich, Horkheimer, and Adorno represented the first groping efforts to establish a literary counter public sphere independent of the official publishing industry; a counter public sphere that would go beyond the production of brochures, leaflets, and conference protocols arising from specific situations. It was, however, not so much the specific analysis of late capitalist society that distinguished critical theory from other theories for this intelligentsia of bourgeois origin (examples of this type of analysis would be Horkheimer's postulate of "integral statism" or his study on commodity fetishism in the culture industry). Much more significant for this young, West German intelligentsia was the *existential form* of a mode of thinking free from all organizational and tactical scruples that could be geared to objective needs. This mode of thinking lent itself to that uncompromising moral demand—severed from traditional patterns of self-interpretation—made by individuals who had betrayed their class with their anticapitalist protest and were now looking for a new, *political* identity. This alone explains the curiosity with which Horkheimer's book *Dämmerung (Twilight)*, first published in 1934 under a pseudonym, was read and appropriated. Taking La Rochefoucauld's *Maximes* as a literary model, Horkheimer, the son of a wealthy industrialist, described and provided reasons for his defection from the bourgeois class.

Horkheimer and Adorno had firmly maintained the position that enlightenment reaches human beings not by appealing to ideals and theories, but by referring to immediate interests and needs. Their work quickly led others to realize that the public sphere is a constitutive category for the experience and consciousness of thoroughly socialized individuals. Jürgen Habermas' *Strukturwandel der Öffentlichkeit (The Structural Change of the Public Sphere)*, published in 1962, was crucial for all attempts on the part of the protest movement to integrate the political contents of critical theory in collective forms within the counter public sphere. Up to that point critical theory was stamped by its anti-authoritarian perspective but was still oriented primarily towards individual emancipation. Habermas' book marks a theoretical as well as a practical turning point in the Left's deliberations on mass media. Once the Left places the Frankfurt School's approach to media theory, which was closely interwoven with culture criticism, into the categorical context of empirical social analysis, critical theory provided a practical political impetus in the strategy for creating a *public sphere* as it was later formulated by the protest movement. But even in Habermas, for whom mass media still represented an integral part of the whole complex of the changing bourgeois public sphere, we find characteristic reductions. These reductions are rooted in their own undetected dependence on the very object of his criticism. For one, he confines his study to the dominant traits of *one* historical form of the public sphere, i.e., the bourgeois public sphere. In comparison with the compact and refined force of the bourgeois public sphere, all forms of a counter public sphere—the plebian as well as the proletarian—appear mainly as repressed variations of the dominant form, cropping up occasionally and disappearing again. Furthermore, this study of structural changes in the liberal model of the bourgeois public sphere fails completely to address itself to that phase of the capitalist social order in which the essence of the bourgeois public sphere manifested itself practically and graphically—Fascism. This was all the more remarkable for a work written within the framework of the Frankfurt School, inasmuch as it was precisely the critical, intellectual coming to terms with Fascism that constituted the political motivation for the analysis of mass media by Horkheimer, Adorno, and their colleagues from the *Zeitschrift für Sozialforschung*.

However, even the protest movement did not overcome the political indecision and theoretical ambivalence to be found in this fundamental analysis of the public sphere by a leftist theoretician. The protest movement developed long-range goals and platforms, initiated media campaigns (as, for example, the anti-Springer movement), created approaches to a new mode of production that brought forth important texts for schooling, books, newspapers, and brochures, and founded its own publishing companies and distribution systems. All of this makes reference to an autonomous *proletarian public sphere*, which, however, did not reach the masses since they were excluded from the start by this mode of production stamped predominantly by intellectuals.

Enzensberger was therefore right in his 1970 retrospective concept of a media theory which sharply criticized all attempts to set up a new mode of production based on craftsmanship in juxtaposition to the *existing* means of mass communication, instead of exploiting the latter's internal contradictions. He poignantly formulated an alternative—a media politics which would quietly undermine the restrictive conditions of the bourgeois public sphere: "Every socialist media strategy must seek to eliminate the isolation of the individual participant in the social learning and production process. That is not possible if those involved do not organize themselves, and this is the political core of the media question. It is on this point that socialist conceptions diverge from late liberal and technocratic ones."[5] Yet, as clear and as sensible as this alternative may appear, the protest movements have had practical experience with the institutions of the bourgeois public sphere that are far more differentiated. The conflicting nature of this experience can scarcely be eliminated by determined strategic action. At most, this is a task for the arduous and contradictory process of developing an autonomous proletarian public sphere.

If the intelligentsia lacked refined strategies for influencing the highly developed mass media, it must first be stressed that this cannot at all be explained by diagnosing the intelligentsia as suffering from a psychological fear of contact and hampered by elitist cultural prejudices against "the masses" and technology. On the contrary, the mass media had to appear to the intelligentsia as monolithic blocks, operating entirely in the interest of capital since the bulk of their operations lay, in effect, in the daily mobilization of prejudices against the anti-capitalist protest movement which were already present in the populace and were

---

[5]Hans Magnus Enzensberger, "Baukasten zu einer Theorie der Medien," in *Massenkommunikationsforschung 2: Konsumtion*, pp. 420–33.

constantly recreated. Nevertheless, the intelligentsia had to acknowledge the fact that the only way for reports of mass demonstrations and strikes, revolutionary slogans, and interviews with speakers for the opposition movement to be given proportionally broad coverage was to use the channels in the technologically highly developed consciousness industry. And this incongruity was an indication of the concrete social situation, i.e., of the *split and polarization in the bourgeois public sphere*. In late capitalism there is an increased breakdown of the traditional public sphere, which continues to draw its self-concept from certain past revolutionary and liberal demands of the bourgeoisie, even though in reality it has long since been reduced to the mere *distribution* of opinions as a result of the emergence of the powerful *public spheres of production* controlled by large conglomerates, the media concentration, and the media systems themselves.[6] These public spheres of production are rooted in the production process itself. Since they are intertwined with or even consist of the technologically highly developed mass media, they mobilize the traditional bourgeois public sphere in order to guarantee particular economic interests and to create mass loyalty for the preservation of the entire capitalist system. Inasmuch as these public spheres pass off particular interests as general ones in the way every ruling class habitually does and, yet, lack any foundation in historical claims to emancipation, it was not at all surprising that the protest movements, which induced people to organize their own experiences and needs, pointed in a demonstrative way to the emancipatory promise of the bourgeois public sphere (as opposed to the secretive practice of those in power) and favored the public sphere of theaters, newspapers, and assemblies in their own work, i.e., communications media still in the tradition of the *citoyen*. In this respect the fact that rebellious French students of May '68 occupied an institution which was rich in tradition like the Odéon theater instead of the radio station must certainly be condemned as politically careless. However, under no circumstances can

6 "The predominant interpretations of the concept of the public sphere are striking in that they try to comprise a variety of phenomena, and, yet, exclude the two most important spheres of life: the entire industrial mechanism of the plant and the socialization in the family. According to these interpretations, the public sphere draws its substance from an intermediary area that does not express any particular social life context in any particular way, although the function of representing the sum total of all social life contexts is attributed to this public sphere." Oskar Negt and Alexander Kluge, *Öffentlichkeit und Erfahrung* (Frankfurt am Main: Suhrkamp Verlag, 1972), p. 10. Today there is a decisive change occurring with regard to this point. The traditional public sphere, the characteristic weakness of which rests in the mechanism of exclusion between the public and the private, is now being overlapped by industrialized public spheres of production. These have the tendency to involve private spheres, in particular the production process and the daily life context, and to utilize them as raw material for secondary exploitation.

the motivation for the occupation be attributed to a clandestine longing for preindustrial conditions.

To be sure, all attempts to appeal to the powerless and partially corroded liberal freedoms and to play them off against the more real and powerful public spheres of production were bound to fail.[7] This resulted in the establishment of a mode of production mainly stressing craftsmanship underneath the communications system. Here the technical means employed in book production, writing posters, formulating leaflets for mass rallies, etc. lagged far behind capital's actual development of the forces of production. Yet, because of that, they could be used as aids, as instruments which one could master. They no longer imposed themselves on the individual in the fetishized shape of self-governing systems. It is obvious that precisely this mode of production, which did not address itself to an abstract audience, i.e., to listeners, watchers, or readers, is suited to a strategy that aims at the politicization of direct experience, above all in the spheres of education and work, and furthers development of autonomous possibilities which would allow people to express their own needs and interests, such as they are articulated in wildcat strikes and other demonstrations, the refusal to work, grass roots movements, and mass demonstrations.

If the subjective richness of human sensuality lying reified and deprived of expression in objects is to be made accessible via experiential appropriation, then the technical aids of generalization and refined cultivation of individual experiences, separate today from direct communication, must again be placed at the collective disposal of human beings. The return to a mode of production based on craftsmanship is the

---

[7]This position appears to be the guiding light behind the official line of the DKP (German CP) in mass communications research today, a position which does not even acknowledge the difference between the traditional public sphere and the public sphere of production. See above all Horst Holzer's writings. "The crucial problem to be treated by the work at hand can be formulated in the question: what do the mass media, newspaper, radio, and television contribute to a democratic society, as described in the basic law of the Federal Republic of Germany?," Horst Holzer, *Massenkommunikation und Demokratie in der Bundesrepublik Deutschland* (Opladen: Leske Verlag, 1969), p. 5. Holzer pursues this line of questioning on three levels:

1. How do the qualities of the mass media which are indirectly demanded by the constitution stand in proportion to what they actually have to offer?
2. What influence do the media's contents have on their audience; in other words, audience's attitudes and reactions?
3. It is important to clarify the specific qualities of the mass media described and the results attained by them with reference to the Federal Republik's particular political-economic structure.

These lines of questioning also determine extensively Horst Holzer's *Gescheiterte Aufklärung? Politik, Ökonomie und Kommunikation in der Bundesrepublik Deutschland* (Munich: R. Piper Verlag, 1971).

futile but understandable attempt to prevent human beings from being robbed of—aside from the economic means of production—their subjective means of expression. In order to counteract this imperceptible but extremely consequential process of deprivation, it is important to use the technical means of communication which are already manufactured in mass production today and available to every wage-earner not only as a private means of production in leisure time, but also as social means of production at work. This is necessary because there is yet another element in the contradiction between a capitalist-industrial mode of production and a mode of production of experiences based on craftsmanship. Inasmuch as mass media force their way persistently into the lives of human beings and try more and more to reorganize them on a capitalist basis, they create the appearance of universality, completeness, and immediacy of information. In so doing they reduce everyday communication to something subjective and random, inconsequential and reserved for leisure time—a fact that influences the narrowly private usage of those means of communication which, like tape recorders, picture and home movie cameras, can be accessible to every individual. Without prior collective, political self-organization, that is, without the possibility of articulating experiences within the framework of a proletarian public sphere, one must even question the emancipatory usage of those instruments of communication which have not been totally appropriated by state and economic monopolies and which can be freely acquired due to mass production.

There is no doubt that it has become necessary to demolish the network of mediations in the bourgeois public sphere and to produce a new immediacy in a demonstrative way through a mode of production based on craftsmanship, for instance, through the creation of autonomous work collectives in publishing companies. However, this step cannot be a substitute for the protracted unfolding of the *dialectic between the bourgeois and the proletarian public spheres*. Every attempt to skirt this is doomed to failure. By eliminating a social area relevant to the consciousness and behavior of human beings such as the field of mass media from its political strategy, the radical protest movement unconsciously reiterated one of the pivotal mechanisms of the bourgeois public sphere— that of persistent exclusion. The renunciation of social theory that could comprehend this dependency and shed light on practical prospects for overcoming it was characteristic of the protest movement and made radicals the victims of that dialectic. Politically, this meant that those forms of a counter public sphere became threatened by decay[8]—a situation which currently confronts the West German Left.

[8]At the present time there is a counter movement that aims to put an end to the theory

III

We can speak meaningfully of critical communications research only if its line of questioning is gleaned from the context of a *theory of society*. Such a theory alone is in a position to diagnose the entire, historical state of society and to deal with the present as a historical problem. The truth content of such a theory is determined above all by the manner in which it succeeds in lending a conceptual voice to social experience. The more specific the historical contents comprised by such a theory are, the more capable of generalization is that theory. "Critical theory" is that form of Marxist theory whose experiential content is determined by Fascism, the highly civilized reversion to barbarity, and which holds firmly, up through the formulation of a "negative dialectic," to the idea that late capitalism in its very dynamic core is potential Fascism. This limitation of theory to the effectual content of a particular historical situation may appear as a constriction of Marx's theory. In reality it is an enrichment, a historical specification that must be performed anew, often unknowingly, by every generation if it is to renew and vitalize the revolutionary critical content of Marx's epoch-making theory.

Fascism had already been established in Germany when the Frankfurt School began publishing essays on mass media in the *Zeitschrift für Sozialforschung* (with the exception of two essays which were published in 1932). The phase was over in which the technologically highly developed mass media, above all radio and film, could still be regarded in terms of the open-ended prospect of stabilizing domination and of enlightenment in the bourgeois and proletarian public spheres. In order to designate more precisely that situation with which the Frankfurt theoreticians were confronted, I shall briefly treat the viewpoints of three authors as exemplary for this beginning phase in the development of the media: Bertolt Brecht, Béla Balázs, and Walter Benjamin.

Brecht's radio theory, consisting of critical notes on a radio play, a speech on the function of radio, suggestions for the radio director and recollections—altogether not more than fifteen pages, is without doubt one of the most important things ever written on the media. Caustically ironic, Brecht denounced technology's colossal triumph in "finally being able to make a Viennese waltz and a cooking recipe eternally available to

---

deficit. It is, however, still stuck in the abstractions of the beginning. It is a question of applying the categories of the critique of political economy to media research. In contrast to the condition of the lack of theory in which the leftists' confrontation with mass media problems took place until now, these attempts undoubtedly mark great progress. Frank Dröge's work, rich in material, may be cited here as exemplary: *Wissen ohne Bewusstsein— Materialien zur Medienanalyse* (Frankfurt am Main: Fischer Verlag, 1972). In reading this work however, one has the suspicion that it falls into Prokop's category of "abstract-critical communications research." Treatment of any kind of a counter public sphere is lacking.

the entire world." In effect he anticipated all the daily nonsense that has since permeated the medium and notes even here, as in the case of the cities which have become uninhabitable, the consoling gesture of the bourgeoisie, which assuages its anxieties about the bad results of a social institution with the line about making its tremendous possibilities. Nonetheless he saw the radio as epoch making. It appeared to him an old contrivance, almost a "mouthpiece" for humans, once lost in the deluge and now rediscovered. Just as for Marx the development of the five senses was the product of all history up to the present, Brecht wanted to reveal even in the beginning stage of radio history the conditions under which the radio could be made into a productive process for changing relationships among people. He wanted to do this in order to obviate the atrophy of the new mass organ for sensual enrichment of human relationships to reality, an atrophy which was clearly perceptible under bourgeois class domination. Brecht's practical suggestions operated on two levels: if the radio remains a mechanism of distribution, then it must publicly expose and denounce those in power in order "to make current events productive." Regardless of whether it is parliamentary sessions, court cases, interviews, or radio plays, that which is reproduced true to reality serves the public education of the masses.

Yet, the actual core of Brecht's radio theory concerned—and was in this respect paradigmatic for every conceivable critical communications research—the transformation of the medium from a *mechanism for distribution of events and opinions into an autonomous production process of experiences:* "The radio must be transformed from a mechanism of distribution into one of communication. The radio would be the most fantastic mechanism of communication imaginable in public life, a tremendous channel system. That is, it would be that if it realized the capacity not only to broadcast, but also to receive; not only to make the listeners hear, but also to make them speak; not to isolate them, but to put them into contact with each other. The radio would thus have to abandon its status as supplier and see to it that the hearer assume that status. For this reason all of the radio's efforts to imbue public affairs with a real sense of the public sphere are absolutely positive."[9] Brecht was aware of the fact that these suggestions, for him merely the natural consequence of the media's technological development, were not practicable in the existing social order. However, he insisted that they can lose their abstractly utopian character if one proceeds "to counter the powers of exclusion with an organization of the excluded," to make the media gradually serve the propagation and formulation of a new social order.

[9]Bertolt Brecht, "Radiotheorie," in *Gesammelte Werke*, 18 (Frankfurt am Main: Suhrkamp Verlag, 1967), p. 129.

Brecht's tendency to regard the new medium as something ante-diluvian and rediscovered, as humankind's socially extended arm, was defined in more detail by Béla Balázs, director of the literature division in the office of the commissioner for the people's education under Georg Lukács during the Hungarian soviet republic. Balázs concentrated on film as an art of seeing, of concretization, in which the photographic close-up technique serves to establish relentless presence and to pierce the murderous abstraction that trims human beings and things in capitalist commodity production. In 1923, when the first radio station went into operation in Germany and when film consisted of not much more than *Kintopp-Kolportage*, Balázs had written: "A really new art would be like a new sense organ," and in 1930 he added: "Since then film has become that. A new organ for human beings to experience the world, and one that has developed rapidly . . . that is more important than the aesthetic value of the individual works which owe their genesis to this organ."[10]

Balázs not only attributes an emancipatory expansion of sensuality to the film by virtue of its inherent logic and fantasy in the unfolding of optical thought, but he also saw film as generating a mass culture independent of the monopoly of the ruling class and in accordance with this new technology of expression and communication. However, he did not dare question the separation of the senses itself as this separation had developed historically. With an eye on the Soviet Union's heroic post-revolutionary periods, Balázs saw in the rise of the sound-film, in the inundation of film with ideologically manipulable language, the danger of destroying precisely that immediacy of concretization that resists capitalist value abstractions. Thus, the members of the petit bourgeoisie class, who have no class consciousness to call their own, but whose mentality is borne by all other classes, appeared to Balázs in the emergence of the sound-film as the prototypes for the future addresses of film—a conjecture which occurred to Brecht when he watched Eisenstein's *Potemkin*. The indignation of the sailors on the battleship moved the audience with such immediacy because the males among them were able to imagine how they themselves might react, were their wives to set rotten meat before them.

Structures of prejudice and ideology of subjectivity, as they dis-tinguish the petit bourgeois mentality in particular, point to those contents later set in motion by Fascism to appropriate the mass media in the interest of maintaining bourgeois class society. This was recognized by Walter Benjamin, who published some essays in the *Zeitschrift für Sozialforschung* and was close to Horkheimer and Adorno. In many

---

[10]Béla Balázs, *Der Geist des Films* (Frankfurt am Main: Makol-Verlag, 1972).

theoretical questions, including aesthetics, an area in which he was closer to Brecht than to the Frankfurt School, he went his own way, as can be seen in his essay "The Work of Art in the Age of Mechanical Reproduction," written after the triumph of Fascism. It was here that he made a last great attempt to glean categories from the demythicizing tendency in the development of the productive forces of the modern means of communication. These categories were meant to deny access to Fascist utilization and to found an alternative socialist policy of art. Benjamin did not want to formulate theses on the art of the proletariat *after* the seizure of power or in a classless society, but rather was concerned with aspects that deal with the current conditions of production. These aspects "brush aside a number of outmoded concepts, such as creativity and genius, eternal value and mystery—concepts whose uncontrolled (and at present almost uncontrollable) application would lead to a processing of data in the Fascist sense. The concepts which are introduced into the theory of art in what follows differ from the more familiar terms in that they are completely useless for the purposes of Fascism. They are, on the other hand, useful for the formulation of revolutionary demands in the politics of art."[11] Benjamin's hopes for the emancipation of art from its parasitic existence in ritual were based on the destruction of aura, of the uniqueness of the here and now, of the measure of genuineness in the production of art, and he saw this destruction as being effected by the technical reproducibility of the ritual. This process marked a revolution in the entire production of art—from being founded in ritual to being founded in politics. However, Benjamin sensed very clearly that the liberating disintegration of the parasitic culture ritual would certainly find a substitute in Fascism. Benjamin's media analysis was oriented towards determining prospects for the *politicization of aesthetics*. In order to help the proletarianized masses along to an expression devoid of and alienated from their objective interests, Fascism made use of *the same media* for a totally different purpose, namely the *aestheticization of politics*. Without film's and radio's means of portrayal, capable of mass influence, Marinetti's apotheosis of war as something beautiful and creative, as he depicted it in his manifesto on the Ethiopian colonial war, would have been unthinkable. To be sure, the aura of the genuine, the unique, was destroyed by mechanical reproduction, but a new aura arose simultaneously: secondary rituals involving the cult of the Führer and the star; images and archetypes which, because they were rooted in prejudices, were all the more suitable for the effective deception of the masses.

---

[11]Walter Benjamin, "The Work of Art in the Age of Mechanical Reproduction," in *Illuminations*, trans. Harry Zohn (New York: Harcourt, Brace, 1968), p. 220.

IV

A proletarian public sphere founded on the beginnings of autonomous class politics, albeit empirically not lending itself at all to a clear definition or localization, say, in class organizations, was the presupposition shared by Brecht, Balázs, and Benjamin. They took it as their point of departure to fuse the intelligentsia's collective mode of production fostered by advanced forces of production in means of communication with the specific organizational forms of experience emanating from the proletariat. For the authors of the *Zeitschrift für Sozialforschung*, however, this connection between the mass media and proletarian public sphere had already collapsed completely. For them the media operate as closed and total institutions and appear as such whether in the guise of Fascist propaganda mechanisms or that of large commercialized operations. Essentially the media are determined to universalize commodity production. In this process the entire cultural heritage is drawn into the capitalist nexus of utilization and domination. All this tends to lead to a critique of ideology in that the illusion of substance created by bourgeois culture is definitely destroyed. Yet, the individuals who are open to such "enlightenment" and could convert it into liberating praxis are themselves confined to socio-psychological dispositions, to behavioral modes characterized by regression.

If, as Horkheimer has claimed, communication and authentic experience have been separated so that the media can neither produce nor mediate experience, then the problem of the public sphere assumes a new dimension. The theoreticians of the Frankfurt School were faced with the disintegration of the bourgeois public sphere and the daily life context in which it was rooted; they realized it was impossible to bestow historical reality on their analyses by identifying with the proletarian mass organizations or the Stalinized Soviet Union.[12] It was characteristic of the development of the Frankfurt School that in light of these things it resorted reflectively to *the experience of the intelligentsia's own mode of production* as the only procedure not requiring any external legitimation. Since theory was the productive format specific to the intelligentsia, it itself became the autonomous basis for formulating anew the intelligentsia's relationship to praxis as well as to empirical fact. Because Fascism was perceived as the epoch-making collapse of the entire bourgeois world, and because on the other hand real developmental trends towards

---

[12]None of the authors writing for the *Zeitschrift für Sozialforschung* were willing to follow in Bloch's footsteps. In the afterword to Bloch's political essays from 1934–39, published in 1972 by Suhrkamp with the title *Vom Hasard zur Katastrophe*, I have tried to root Bloch's conflicting position on Stalinism by locating it in the situation of intellectuals at that time. This essay appeared in *New German Critique*, 4 (1975), 3–16, under the title "Ernst Bloch— The German Philosopher of the October Revolution."

proletarian emancipation were no longer discernible, the proletariat was transposed into the *idea* of the proletarian public sphere and made into the object and content of a mode of production which claimed its legitimation in collective emancipatory interests, but which was still determined individualistically in the concrete work process. The radical return of theory to the core of its own experience meant that the traditional cognitive modes and categories revealed in their oblivious objectivism a part of the very catastrophe they presumed to comprehend. Ideas and behavioral modes that point beyond the given organization of existence, the promise of freedom and happiness, always include traits of affirmative culture, the camouflaging of daily misery, domination, and exploitation. A reconstruction of the generic-historical dialectic inherent in culture that might reveal this would therefore be constitutive of a theory meant to deal with the present on its own terms. Given this, every authentic form of production on the part of the intelligentsia—whether it be the sociologist, the philosopher, or the artist—takes on an epistemological character.

Although Marx's critique of political economy (from which they drew their concept of criticism to begin with) remained the basis for their explanation of Fascism, Horkheimer, Adorno, and the social scientists grouped loosely around the *Zeitschrift für Sozialforschung* made an essential contribution to the explanation of Fascism in their comprehensive, empirically sound, and psychoanalytically oriented studies on authority and family, authoritarian character structures, etc.[13] Adorno, as director of the musical division of the Princeton radio research project, even studied empirically the radio's functional mechanisms with regard to its neutralizing musical content and to the tendencies furthering regression of the human senses. However, these studies are guided to such an extent by the dialectical social theory specific to the Frankfurt School that it will suffice here to treat two problem areas that precisely designate the direction of the critical cognitive interest—the social status of the critical intelligentsia and the function of art—both primarily from the standpoint of production.

Proceeding from Valéry's concept of construction as developed in

[13]In this context a brief evaluation of critical theory's relationship to psychological research is necessary. Critical theory acknowledges the dialectic between economic base and critical consciousness. It counters the enlightenment critical solely of economy with a system that includes psychoanalytic, social, and aesthetic elements. This must be taken as a point of departure by every criticism that intends a change in human beings' consciousness, their psychology, their needs and hopes. The psychoanalytic element is present in the *Zeitschrift für Sozialforschung* and represented above all by Erich Fromm and Theodor W. Adorno. It is of essential significance. Nevertheless, it is necessary to interpret this regard for psychoanalytic categories as a political problem in order to counteract the suspicion of psychologizing social states of affairs from the start.

Monsieur Teste, Walter Benjamin considered the modern author's view of his/her own working and procedural mode decisive for determining the French writer's current position. This reflexive turn to the production process, the sublation of the unconscious reified production that had always been able to find a basis in bourgeois cultural norms, even if its products were in effect already being adapted to the machinery of utilization, was in fact the first act of liberation by the intellectual producers and the first act of redefinition of the producer's truth content. Benjamin was the one who described the critical intellectual's ambivalent position most clearly and most poignantly: intellectuals see the bourgeois world break down and cannot reconcile the empirical condition of the revolutionary workers' movement with their own conceptions of revolution. This is due to the fact that the movement is marked by defeats, demolition of collective forms of organization, bureaucratic deformation, and imprisonment of the avant garde, and the intellectuals would like to pour all they have learned into this movement with great subjective exuberance. If French surrealism saw itself as "winning over the forces of intoxication to the revolution," this was true in modified form for the other writers and philosophers who abandoned their class. Benjamin's insight is extremely meaningful today: "Just because this intelligentsia has forsaken its class in order to make the cause of the proletarian class its own does not mean that the latter has accepted the former. It has not. Hence, the dialectic in which Malraux's heroes operate. They live for the proletariat; yet, they do not act as proletarians. At least, they act much less out of class consciousness than out of the consciousness of their loneliness. That is the torment from which none of them can escape. It also constitutes their dignity."[14]

During the French Revolution Kant lauded the solemnity in which law was affirmed and a constitution in accordance with natural rights was evolved. The situation of the circle of intellectuals aroung the *Zeitschrift für Sozialforschung* was similar to Kant's, though they were influenced more by social circumstances than their bourgeois origins. In all probability none of them participated actively in the organizational work of a proletarian party or union, but they all wanted to salvage the historical-substantive forces of reflection for the revolution in philosophy as well as in culture. Inasmuch as they saw themselves, not as the avant garde of these forces, but as their conscience, the standard criticism of them as lacking praxis misses the mark and loses sight of *their* concept of the relationship between theory and praxis. The unification of contradictions embedded in this relationship cannot be consistently maintained in every

---

[14]Walter Benjamin, "Zum gegenwärtigen gesellschaftlichen Standort des französischen Schriftstellers," in *Kritische Kommunikationsforschung*, p. 224.

situation but is itself subject to a historical dialectic: "Intellectuals who proclaim and worship the proletariat's creativity and are content to follow and glorify the proletariat fail to see that they make the masses blinder and weaker than is necessary. This happens each time that intellectuals avoid straining themselves theoretically so that they become passive and circumvent a temporary opposition to the masses which their own thought might engender. As a critical, propelling element, the thoughts of intellectuals belong to the development of the masses."[15]

The proletarian public sphere, in which subjective interests and needs, experiences and ideological inversions are articulated and assume physical force, may remain the frame of reference for critical theory's reflections in the 1930s. Yet, even here we find a barely perceptible but decisive shift. The objective interests which are rooted in the given developmental stage of the social forces of production and determine what standard of emancipation is to be regarded as necessary are removed from the real historical movement, from the social process of emancipation of the working class, and transposed into theory. Indeed, Horkheimer stressed time and again that the decision in favor of revolutionary, liberalist, or Fascist aims and means is not the result of apparently neutral reflection, but a question of political struggle, of concrete historical activity: "The avant-garde needs shrewdness in political struggle, not academic instruction on its so-called point of view." But his anti-academic affectation, identifiable throughout the Frankfurt School, is not sufficient to overcome the level of that organization of experience which is solely characteristic of *theory's mode of production*.

Horkheimer understood theory as a class-conditioned mode of organizing social experience. Traditional theory proceeds from a subject-object dualism prescribed by natural science's cognitive model. Traditional theory does so because it considers it logically binding that the changing of objects under investigation, affected consciously by research but not lending itself completely to checks, makes a "regulated transformation of communicative experiences into data" and, thus, measurements are impossible.[16] Critical theory does not, in any case, understand its relationship to empirically comprehensible reality as a variation of action research. It makes use of advanced methodological tools: it uses interviews and multi-faceted computation procedures; it constructs hypotheses and describes places of work. In so doing it simultaneously dissolves the discipline's reified framework and sees itself as a product of experience, organized in a particular fashion—as a selection from and a

---

[15]Max Horkheimer, *Traditionelle und kritische Theorie: Vier Aufsätze* (Frankfurt am Main: Fischer Verlag, 1975), p. 33.

[16]Jürgen Habermas, *Theorie und Praxis* (Frankfurt am Main: Suhrkamp Verlag, 1971), p. 18.

constitutive part of social processes. The truth to which critical theory lays claim is not given in the objects themselves. It is not, as in bourgeois science, the mere conformity of the scientific image with the analyzed facts, or in other words, according to the traditional criterion of truth, the *adaequatio intellectus atque rei*. The truth to which critical theory lays claim establishes itself only in the process of historical change: it is truth as process. Factual statements and normative outlooks do not diverge from each other, but rather constitute in every individual cognitive step this dialectical concept of truth.

Since, logically speaking, this concept necessarily involves the theoretician's objective involvement in the struggle for emancipation, it is by no means enough to speak consistently from the proletarian class standpoint in order to think correctly. Much more, the materialist science linked to a certain productive and commercial mode of experience is drawn in accordance with its form and content into the context of proletarian life, in which context the experiences of the masses are organized in a particular manner. In this context of proletarian life the question is one of empirical relationships, i.e., of the worker's material and psychological dependencies on capitalist commodity production, which reifies all human relationships. Thus, the context of proletarian life is at the same time characterized by an *obstruction of experiences*. Neither Horkheimer nor the Frankfurt School advanced beyond academic theory. The proletarian public sphere detached from the real daily life context of the working class became an idea serving as a depository for and a reminder of everything having emancipatory content in human history and now threatened with extinction.

Trotsky designated Fascism as a particular state system whose task was not only to destroy the proletarian avant garde, but, above all, to annihilate all elements of proletarian democracy in bourgeois society and to keep the entire class in a state of induced dissipation. It is then clear that the truth content of theory can be founded neither in the vague hope of socialism's final victory nor in the ascertainment of the proletariat's empirical constitution. Horkheimer was right when he said: "Even the systematization of the proletariat's consciousness would not be able to provide a true picture of its existence and its interests. It would be traditional theory with a particular focus, not the intellectual side of the historical process of the proletariat's emancipation." Neither the demand that the Marxist theoretician be true to the need for emancipation inherent in every human being, nor the necessary insistence on objective historical interests, nor the "persistence of fantasy" among the most advanced groups in the working class can alleviate the problem posed by critical theory in its reformulation of the dialectical relationship between theory and praxis, subject and object, theoretical and empirical analysis.

Critical theory posed this problem but was not capable of solving it because critical theory contained elements which tended toward a reversion to traditional theory in its very origins.

Critical theory originally conceived itself as the intellectual side of the real revolutionary emancipatory process. The fact that it did not reach the masses, that it not build a material base of power, may be rooted in the unfavorable circumstances of a historical situation, but at the same time it affected its truth content. Indeed, it is essential to continue reflecting on *how* scientific organizational forms of experience and the modes of experience that occur in proletarian daily life (for example, production of fantasy as the masses' authentic cultivation of experience) are and must be linked with one another in order to further organization and the collective unfolding of the individual's social possibilities and needs. It *is* the dialectical designation of the proletarian class standpoint. Only by locating science and theory in a continuum of experience can we be certain that proletarian contents can penetrate science and conversely that scientific results can constructively enter the proletarian process of experience.

Where critical theory saw the possibility for authentic experience in the production sphere, this experience was marked by absence of communication. The Frankfurt School's avant-garde oriented cultural theory did not originate in the mere fact that its representatives were primarily from the bourgeoisie; it was, rather, mediated by the social situation, the result of political desperation and hopelessness. The crucial significance that music attained under these circumstances was due to the fact that music allows for the broadest spectrum of expression and the most acute potential for communication without the risk of concealing and mediating this verbally or ideologically. This is the case with the uncompromising atonality of the Schoenberg School, the most advanced musical production, as well as the mass production of hit songs, which are more readily understandable than the simplest dialogue: "Nowadays art no longer relies on communication."[17] It is not only a question of mourning the vanishing bourgeois individuality (H.J. Krahl), but of saving a past life and anticipating an objectively possible better one—a result of experiencing a historical situation that did not seem to offer any other way out: "The work of art is the only fitting objectification of the individual's destitution and desperation."[18] Yet, if this is true, then art— to the extent that it, in consideration of the "aesthetic difference," as unconscious writing of history, lends representation to social relationships in accordance with its own material and structural laws—shares the

[17]Horkheimer, "Neue Kunst und Massenkultur," in *Kritische Kommunikationsforschung*, p. 33.
[18]Horkheimer, "Neue Kunst und Massencultur," in *Kritische Kommunikationsforschung*, p. 34.

problems of a dialectical social theory. Its main function is that of cognition: "If the most advanced musical production of the present puts fundamental bourgeois categories, such as the creative personality and its expression of soul, the world of private feelings, and glorified subjectivity, out of commission and puts in their stead highly rational, transparent constructive principles, this music, bound to the bourgeois production process, is certainly not to be regarded as 'classless' and the real music of the future. It is, however, to be regarded as that music which fulfills its dialectical cognitive function with greatest precision."[19] However, the function of aesthetic production to be critical of ideology and the salvation of human possibilities for experience and communication can be realized only at the price of isolation, the break with reified and speechless communication, by means of which the masses are kept at the level of lack of experience. Thus, as Adorno showed, music's isolation cannot be corrected simply in terms of music itself, but solely by changing the entire state of society.

In contrast to this, every form of the classicist return to the past proves to be a deceptive cover-up for inhuman conditions and signifies a poetic appropriation of the past. It is only in those areas in which there is a living cultural mode of production following autonomous laws that it seems possible to keep the cultural heritage intact. The dialectical notion of "the sublation of the cultural heritage" (die Aufhebung des kulturellen Erbes) implies the realization of both negativity and preservation. However, under existing conditions, the emphasis seems to shift toward preservation. For most theoreticians of the Frankfurt School this area of living cultural production was artistic production, in particular music and literature. Advanced art alone preserves the utopia, which was once a structural element of the entire culture.

If a living productive mode of experience is the prerequisite for the appropriation and preservation of past culture, this salvation, which aims at the formation of historical consciousness, can succeed only if it enters into the public context of the living mode of production of the masses' experiences and consistently bursts the framework of the bourgeois public sphere. The Frankfurt School, confined to the level of theory production, formulated beginnings for such a transformation process. Here lies the substance of critical theory. Those familiar with Horkheimer's biography know that his critical thought ended in a religious and anti-Marxist phase. Despite the subjective turns of some members of the Frankfurt School, however, and despite the positivism that characterizes

---

[19]Theodor W. Adorno, "Zur gesellschaftlichen Lage der Musik," in *Kritische Kommunikationsforschung*, p. 156.

some of their successors, the productive aspects of critical theory deserve to be upheld.

Translated by Leslie Adelson
Washington University

The Instrumentalization of Fantasy: Fairy Tales and the Mass Media

JACK ZIPES

Ever since the eighteenth century German bourgeois writers have shown a marked propensity for writing and studying folk and fairy tales. The rich vein of culture tapped by this activity, and still being explored today, is vital to a concrete realization of humanistic utopian projects. Instituted by Herder, whose ideas inspired other writers of the period such as Goethe and the poets and dramatists of the Storm and Stress Movement, interest in folk and fairy tales persisted among the Romantics, leading Novalis to proclaim that: "The genuine fairy tale must be at the same time a prophetic portrayal—[an] ideal portrayal—[an] absolutely necessary portrayal. The genuine fairy tale writer is a seer. (With time history must become a fairy tale—becoming once again what it was in the beginning.)"[1]

The notion of history as fairy tale might easily be dismissed as romantic escapism; on the contrary, however, the utopian landscape of Novalis' fairy tales is a rebellious response to reason's having been pressed into the service of authoritarian powers, the Enlightenment having rationalized human production for economic exploitation and profit, thus warping the people's sense of their own history. For Novalis, history was a reappropriation of nature and human resources by individuals engaged in mutual and harmonious pursuit of their full

---

[1]Novalis, *Werke und Briefe*, ed. Alfred Kelletat (Munich: Winkler Verlag, 1962), p. 506.

potential through love and the imagination. The fairy tale showed the way toward self-realization: the alternative path history could take if human beings actually took charge of their destiny.

Novalis was already fighting a losing battle with his fairy tales and radical theories. Nonetheless, the great German bourgeois writers up to the present have continued to use folk and fairy tales in order to critique the dehumanizing forces of rationalization and capitalism, and while their love of the fantastic may be interpreted as a retreat from reality, it also signifies an active use of the imagination to oppose social manipulation and arbitrary domination. The telling of a folk or fairy tale involves an autonomous exercise of the imagination which endows the teller with a sense of his or her own power and challenges the self-destructive dictates of reason. Equally important is the creator's direct connection with an audience, with a social experience, and with nature. As Walter Benjamin has emphasized, "a great storyteller will always be rooted in the people"[2] since he or she has the practical task of communicating wisdom as a use value to the people, and such mediation can effectively bring audiences closer to nature and endow them with a sense of their potential for self-realization. Benjamin particularly praised the folk tale as the highest form of narrative:

> The folk tale, which to this day is the first tutor of children because it was once the first tutor of mankind, secretly lives on in the story. . . . The folk tale tells us of the earliest arrangements that mankind made to shake off the nightmare which the myth placed upon its chest. . . . The wisest thing—so the folk tale taught mankind in older times, and teaches children to this day—is to meet the forces of the mythical world with cunning and high spirits. . . . The liberating magic which the folk tale has at its disposal does not bring nature into play in a mythical way, but points to its complicity with liberated man.[3]

Like Novalis, who sought to retain the original purpose and authenticity (aura) of the folk tale in his fairy tales, Benjamin sees the folk tale as a quasi-magical mode of connecting the people with their own nature and history. Myth as the social construct of instrumental reason must be pierced and exploded by the powers of the imagination, which maintains a use value by providing counsel to the people and restoring the immediacy of nature to them. The folk tale then was a model of narrative art which, despite its gradual decline in the age of mechanical reproduction, could still serve to point the way toward the making of real history. In the twentieth century this would mean that a tale would have to provide counsel against authoritarianism and commodity fetishism

---

[2] Walter Benjamin, "The Storyteller," in *Illuminations*, trans. Harry Zohn (New York: Harcourt, Brace, 1968), p. 101.

[3] Benjamin, p. 102.

and subvert instrumental rationality by the imagination, as in the following example:

> Once upon a time there was a rich and popular young man who was the delight of his father's employees, the poor young woman whom he had chosen as his fiancée, and the artists and intellectuals whom he had befriended. Everyone found him charming until his father's business went bankrupt, then his charm abruptly began to pall. Not that the young man had remained the same while everyone else had changed, but rather that the collapse of his father's business had given his character an entirely different meaning. He was found to be tiresome, untalented and unproductive; thus a diminishing bank balance is sufficient to render fatuous what was once endearing. The effect would have been even more dramatic had circumstances allowed the others to know about the father's reversal before his son did, thus causing him to become an idiot (in the eyes of others) without the slightest change in his own consciousness. So little are we dependent on ourselves.[4]

This fairy tale (which I have paraphrased) is not the work of a romantic but of Heinrich Regius, a pseudonym for Max Horkheimer, one of the founders of the Institute for Social Research in Frankfurt, who used the fairy tale to demystify the operations of commodity fetishism in capitalist society. While this is not the place for a detailed interpretation of Horkheimer's fairy tale, it is, nonetheless, appropriate to question the potential of the genre as practiced by Horkheimer or the mass media, to lay bare the pernicious effects of the culture industry in order that audiences, unlike Horkheimer's young man, can become *dependent on themselves*.

It is somewhat ironic that Horkheimer and the Frankfurt School of critical theory, whose energies were directed towards recuperating the emancipatory potential of reason, should set the parameters for a critical analysis of the fairy tale and the imagination. The contradiction is merely an apparent one, however, for they themselves relied upon and defended the power of the imagination to subvert the use of reason as an instrument of domination in a society oriented toward the reification of human beings as commodities by authoritarian interest groups. Thus, we can best grasp the contemporary potential of the fairy tale as a liberating cultural force by placing it in the context of what Horkheimer and Theodor Adorno called the culture industry, that is to say, the mode by which cultural forms are produced, organized, and exchanged as commodities so that they no longer relate to the needs and experiences of the people who create them. The notion of the culture industry will be the point of departure for my discussion of the fairy tale. An outline of the

---

[4]Heinrich Regius, *Dämmerung: Notizien in Deutschland* (1934; rpt. Zurich: Edition Max, 1972), p. 102. I have paraphrased the tale for the sake of brevity.

historical development of the folk tale as fairy tale will follow, leading to a critique of Ernst Bloch's theory of the genre's utopian potential, and finally to a reiteration of Novalis's demand that history become a fairy tale, and Horkheimer's implicit claim that we must become dependent on ourselves in order to make history.

I

As early as 1944 Horkheimer and Adorno demonstrated in *Dialectic of Enlightenment* the ways and means by which the culture industry employs technology and instrumentalizes reason to extend the domination of capitalism and make human beings and their cultural expressions into commodities. Concerned about how the Enlightenment itself betrayed the very principles of critical and rational consciousness which it had first championed, Horkheimer and Adorno argued that the organization of the socio-economic system based on private property, competition, and profit was arbitrarily developed through rationalism to prevent human beings from realizing their full imaginative and intellectual potential. They have become little more than tools, it was argued, for while they were required to place their skills and thought at the service of a system which uses industry and technology to increase the profit and power of elite groups, they were prevented from pursuing their own interests and internalized the norms and values of capitalist commodity production. Progress as the advancement of machines and technology for production has become identified with the power of the capitalist system to dominate and manipulate humanity and nature. Both reason and imagination have atrophied. According to Horkheimer and Adorno, culture's role in reinforcing the dehumanization process of capitalism since the eighteenth century has become the key to comprehending how and why this system continues to endure. They were particularly sensitive to totalitarian tendencies in culture, and here they stress that the process by which people and cultural forms are made into commodities through the mass media in the twentieth century has all but destroyed the human ability to distinguish the real from the unreal, the rational from the irrational. Cultural production then, makes people into consumers who out of a sense of confusion and impotence gratefully relinquish their own autonomy. In *Dialectic of Enlightenment,* we read:

> The stunting of the mass-media consumer's powers of imagination and spontaneity does not have to be traced back to any psychological mechanisms, he must ascribe the loss of those attributes to the objective nature of the products themselves, especially to the most characteristic of them, the sound film. They are so designed that quickness, powers of observation, and experience are undeniably needed to apprehend them at all; yet sustained

thought is out of the question if the spectator is not to miss the relentless rush of facts. Even though the effort required for his imagination is semi-automatic, no scope is left for the imagination.[5]

The structure of mass amusement now resembles that of the production line and is intended to lull the masses into believing that the goods they consume and produce actually nourish their potential to develop. Yet, the purpose of production is to make the people into bigger and better consumers with no regard for the quality of the things they produce and consume. Thus, according to Horkheimer and Adorno:

> Amusement under late capitalism is the prolongation of work. It is sought after as an escape from the mechanized work process, and to recruit strength in order to be able to cope with it again. But at the same time mechanization has such power over a man's leisure and happiness . . . that his experiences are inevitably after-images of the work process itself. The ostensible content is merely a faded foreground; what sinks in is the automatic succession of standardized operations. What happens at work, in the factory, or in the office can only be escaped from by approximation to it in one's leisure time. . . . Pleasure hardens into boredom because, if it is to remain pleasure, it must not demand any effort and therefore move rigorously in the worn grooves of association. No independent thinking must be expected from the audience; the product prescribes every reaction: not by its natural structure (which collapses under reflection), but by signals.[6]

The gratification promised by the culture industry turns out to be a dulling of the senses. Since the conditioning images and products of mass media are aimed at making all producers into passive consumers and all art works into commodities, the autonomy of both producer and product is circumscribed by the conditions of the market. Production for consumption and profit become values in and for themselves. Human beings lose touch with their innate abilities and talents and lose sight of the manner in which they project themselves onto the world. The subjectivity of the individual has become so invaded at work and in the home by the demand for standardization that the autonomy of the mind is called into question.

In effect, Adorno and Horkheimer's study of the culture industry posed the major question of our time, a question made particularly acute by the rise of fascism since the 1920s, that is to say: given the extent to which the state and private industry have collaborated to increase their administrative and bureaucratic control over our public *and* private lives through technology, can one actually speak about subjective self-

[5]Theodor Adorno and Max Horkheimer, *Dialectic of Enlightenment* (New York: Seabury Press, 1969), pp. 126–27.
[6]Adorno and Horkheimer, p. 137.

realization and self-determination?[7] Has reason become so instrumentalized for the purpose of capitalist realization that people's inherent
creative and critical potential to shape the destiny of humankind has
become depleted?

Herbert M. Marcuse pursues the question further still, by making
clear how the capitalist ideology of political domination not only pervades
the social organization of human beings but is constitutive of the form and
content of technology itself. Thus,

> The technological a priori is a political a priori inasmuch as the transformation
> of nature involves that of man, and inasmuch as the "man-made creations"
> issue from and reenter a societal ensemble. One may still insist that the
> machinery of the technological universe is "as such" indifferent towards
> political ends—it can revolutionize or retard a society. An electronic
> computer can serve equally in capitalist or socialist administrations; a
> cyclotron can be an equally efficient tool for a war party or a peace party. . . .
> However, when technics becomes the universal form of material production,
> it circumscribes an entire culture; it projects a historical totality—a "world."[8]

The consequences of Marcuse's theory are just as frightening as those of
Adorno's and Horkheimer's: the total saturation of culture by a technology politically and socially geared to curtail critical thinking and
autonomous decision-making leads to a rationally totalitarian society. In
this regard, the mass media function to curb the emancipatory potential
of art and instrumentalize fantasy.

The pessimistic outlook of the major members of the Frankfurt
School has been balanced somewhat by Jürgen Habermas, who makes a
distinction between instrumental action governed by technical rules
based on empirical knowledge and "communicative action or symbolic
interaction, governed by binding consensual norms which define reciprocal expectations about behavior and which must be understood and
recognized by at least two acting subjects."[9] In contrast to Marcuse,
Habermas shifted the focus from technology as the framework within
which capitalist ideology was made manifest and controlled the lives of
individuals, to the institutional framework of the bourgeois public sphere
which determines and embraces the socialization process and technology
composed of systems of instrumental and strategic action. As a historical
sociological category, the public sphere designates those forms of

[7]For a discussion of the Frankfurt School's work in this area, see the chapter entitled
"Aesthetic Theory and the Critique of Mass Culture," in Martin Jay, The Dialectical
Imagination: A History of the Frankfurt School and the Institute of Social Research, 1923–1950
(Boston: Little Brown, 1973), pp. 173–218.

[8]Herbert M. Marcuse, One-Dimensional Man (Boston: Beacon Press, 1964), p. 154.

[9]Jürgen Habermas, "Technology and Science as 'Ideology,'" in Toward a Rational Society
(Boston: Beacon Press, 1970), p. 92.

communicative action developed and institutionalized by the bourgeoisie in the eighteenth century between the private sphere and the state to foster rational discourse and a democratic decision-making process. The causes for the loss of individual autonomy and democracy and the increased technological control of our lives by capitalist interests can only be comprehended, according to Habermas, if one studies how private interests have manipulated and dominated public opinion in the public sphere and how the state has intervened in behalf of these private interests which have developed monopoly power. The result has been a perversion of the public sphere or what Habermas calls a "refeudalization of the public sphere"[10] where the state openly and arbitrarily exercises its control to further the interests of large monopoly and conglomerate concerns. He explains:

> The substitute program prevailing today . . . is aimed exclusively at the functioning of a manipulated system. It eliminates practical questions and therewith precludes discussion about the adoption of standards; the latter could emerge only from a democratic decision-making process. The solution of technical problems is not dependent on public discussion. Rather, public discussions could render problematic the framework within which the tasks of government action present themselves as technical ones. Therefore, the new politics of state interventionism requires a depoliticization of the mass of the population. To the extent that practical questions are eliminated, the public sphere also loses its political function.[11]

Central, then, for Habermas is a theory of communication, not technology, and thus he seeks to overcome the gloom of the Frankfurt School by positing an emancipatory notion of rationalization which is based on "public, unrestricted discussion, free from domination, of the suitability and desirability of action-orienting principles and norms in light of the socio-cultural repercussions of developing subsystems of purposive-rational action."[12] The possibility for repoliticizing the public sphere to allow for democratic decision making and individuation has become the subject of numerous studies in West Germany. The debate among those critics concerned with the culture industry has centered on ways to create a counter public sphere[13] or on ways to expose and oppose the commodity fetishism based on the economic exploitation of the subjective needs of the masses.[14] All these studies up to the present have

[10]See Strukturwandel der Öffentlichkeit (Berlin: Luchterhand, 1962), pp. 157–98.
[11]Habermas, "Technology and Science as 'Ideology,'" p. 103.
[12]Habermas, "Technology and Science as 'Ideology,'" pp. 118–19.
[13]See Oskar Negt/Alexander Kluge, Öffentlichkeit und Erfahrung: Zur Organisationsanalyse von bürgerlicher und proletarischer Öffentlichkeit (Frankfurt am Main: Suhrkamp Verlag, 1972).
[14]See Wolfgang Fritz Haug, Kritik der Warenästhetik (Frankfurt am Main: Suhrkamp Verlag, 1971).

recognized and examined the power of the mass media to increase domination by the culture industry and the threat this poses for the individual consciousness.

The Frankfurt School's critique of the culture industry has not had a profound effect among Anglo-American radical commentators on mass communications, although an influence can be traced through the mediations of Marcuse, Hans Magnus Enzensberger, Raymond Williams, Stanley Aronowitz and Stuart Ewen,[15] whose works have had comparatively wide circulation in the United States and England. The dominant tendency in the U.S. and England in the field of mass communications has been, however, toward empirical and quantitative studies which concretely demonstrate the growing power of monopoly interests without overt reference to the Frankfurt School, while confirming its general thesis. The Vietnam war and the Watergate scandal have stimulated inquiry into the manipulation and machinations of the culture industry. Two works by Herbert I. Schiller and Michael R. Real, respectively, are cases in point.

Schiller's *The Mind Managers* covers the areas of governmental cooperation with private concerns, the military corporate industry, the entertainment sphere, the polling industry, international conglomerates, and the legal system of repression. In each case he proves conclusively that the images and information circulated are not neutral but contain ideological messages which intentionally create a false sense of reality and produce a consciousness that is at the mercy of a commodity industry seeking profit and creating divisiveness and alienation among the people. Real's book focuses on culture: "Mass mediated culture primarily serves the interests of the relatively small political economic power elite that sits atop the social pyramid. . . . For example, while allegedly 'giving the people what they want,' commercial television maximizes private corporate profit, restricts choices, fragments consciousness, and masks alienation."[16] Real elaborates a category of "mass-mediated culture" which can be used to clarify how the culture industry maintains its pervasive influence on mass behavior, and he argues that technological developments, particularly in the age of mass reproduction, have given a new

[15] See Hans Magnus Enzensberger, *The Consciousness Industry*, ed. Michael Roloff (New York: Seabury Press, 1974); Raymond Williams, *The Long Revolution* (New York: Columbia Univ. Press, 1961); Raymond Williams, *Television: Technology and Cultural Form* (New York: Fontana/Collins, 1974); Stanley Aronowitz, *False Promises: The Shaping of American Working-Class Consciousness* (New York: McGraw Hill, 1973); Stuart Ewen, *Captains of Consciousness: Advertising and the Social Roots of the Consumer Culture* (New York: McGraw-Hill, 1976). Williams does not borrow directly from the Frankfurt School, but his work derives from many of the same critical premises.

[16] Michael R. Real, *Mass-Mediated Culture* (Englewood Cliffs, N.J.: Prentice-Hall Inc., 1977), p. xi.

orientation to culture which he defines as the systematic way of construing reality that a people acquires as a consequence of living in a group. Real uses the concept *mass-mediated culture* to argue that "all culture when transmitted by mass media becomes in effect popular culture. Popular culture *not* transmitted by mass media exists but has decreased in importance both aesthetically and socially."[17] Yet, Real, Habermas, Marcuse, and other radical critics have not discarded the hope that culture can still be reappropriated by the masses to serve the emancipatory interests of humantity. But, before discussing the possibility of the radical reorganization of the culture industry, I will turn to the historical development of the fairy tale and the way in which its cultural form and emancipatory content have been changed through technological mass mediations.

II

Any discussion of the fairy tale must take account of its being a mass-mediated form of folk art. Originally the folk tale was, as it remains, to a certain degree, an oral narrative form cultivated by the common people. Originating as far back as the Megalithic Period, the tales are reflections of the social order in a given historical epoch, symbolizing the aspirations, needs, dreams, and wishes of the people, affirming the dominant social values and norms or demonstrating the need to change them.[18] The tales were told by gifted narrators before audiences who actively participated in the transmission by posing questions, suggesting changes, and circulating the tale among themselves.

While folk tales and folk-tale motifs have always served as the basis for more refined and subtle forms of cultural expression, as in Shakespeare's plays,[19] for example, or in the works of Homer and the Greek dramatists, the most interesting aspect of the historical development of the folk tale is its appropriation by aristocratic and bourgeois writers in the sixteenth, seventeenth, and eighteenth centuries, so that with the expansion of publishing, it became a new literary genre: the fairy tale. As a literary text which experimented with and expanded upon the folk tale, the fairy tale reflected a change in values and consequent ideological conflicts during the transitional period from feudalism to early capitalism. All the early anthologies, *Le piacevoli notti* (*Delightful Nights*, 1550) by Giovanni Francesco Straparola, *Pentamerone* (1634/36) by Giambattista

[17]Real, p. 14.
[18]See August Nitschke, *Soziale Ordnungen im Spiegel der Märchen*, 2 vols. (Stuttgart: Fromann, Friedrich Verlag, 1976).
[19]Cf. Max Lüthi, *Shakespeares Dramen*, 2nd. ed. (1957; rpt. Berlin: Schmidt Verlag, 1966) and Robert Weimann, *Shakespeare und die Tradition des Volkstheaters* (Berlin: Aufban Verlag, 1967).

Basile, and *Histoires ou contes du temps passé* (*Stories or Tales of the Past*, 1696/97) by Charles Perrault demonstrate a shift in the narrative perspective and style which not only obliterated the original folk perspective and reinterpreted the experience of the people for them but also promulgated a new ideology. This was most apparent in France during the eighteenth century when fairy tales written by such aristocratic ladies as Madame d'Aulnoy, Madame de Beaumont, Mademoiselle de Heritier, and Madame de Murat were highly appreciated.[20]

Beauty and the Beast admirably illustrates the folk tale's drastically altered nature. Although the transformation of an ugly beast into a savior as a folklore motif can be traced to primitive fertility rites, the theme of this aristocratic fairy tale is "putting the bourgeoisie in their place." It is Madame Leprince de Beaumont's version, published in 1756, which provided the basis for the numerous English translations that have been widely circulated since, and which has effectively corrupted the original meaning of the folk-tale motif while seeking to legitimize the aristocratic standard of living in contrast to the allegedly crass, vulgar values of the emerging bourgeoisie.[21]

The tale concerns a very rich merchant whose children become arrogant because of the family's acquired wealth and, with the exception of Belle, aspire beyond their class. Hence, the family must be punished. The merchant loses his money and social prestige, and all the children are humiliated. Yet, they remain haughty and refuse to help the father, particularly the two older daughters. Only Belle, the youngest, exhibits modesty and unselfishness, and only she can save her father who must lose his life for transgressing against the beast. A model of industry, obedience, humility, and chastity, Belle helps her father, and later, impressed by the noble nature of the beast (appearances are obviously deceiving, i.e., aristocrats may act like beasts but they have gentle hearts), she agrees to give him a kiss and marry him. He is immediately transformed into a handsome prince and explains that he had been condemned to remain a beast until a beautiful virgin should agree to marry him. She is rewarded for having preferred virtue to either wit or beauty while her sisters are punished for their pride, anger, gluttony, and idleness, and are turned into statues to be placed in front of their sister's palace. Surely, this is a warning to all those bourgeois upstarts who forgot their place in society and did not control their ambition.

This tale illustrates the instrumentalization of fantasy, that is to say

[20]See the "Nachwort" by Klaus Hammer to the edition *Französische Feenmärchen des 18. Jahrhunderts* (Berlin: Schmidt Verlag, 1974). Also useful is Gonthier-Louis Fink, *Naissance et apogée du conte merveilleux en Allemagne 1740–1800* (Paris: Belles Lettres, 1967), pp. 11–73.

[21]For more information on the history of *Beauty and the Beast*, see Iona and Peter Opie, *The Classic Fairy Tales* (London: Oxford Univ. Press, 1974), pp. 137–38.

the setting of products of the imagination in a socio-economic context in order to limit the imagination of the receivers. In *Beauty and the Beast*, not only is a folk-tale motif transformed and adorned with Baroque features by the imagination of the writer, but the literary mediation controls the production, distribution, and reception of the tale. As an innovative, privately produced text which depended for its existence on the technological development of printing and the publishing industry, the fairy tale in the eighteenth century excluded the common people and addressed the concerns of the upper classes. Its themes and figures were designed to appeal to and confirm the aesthetic tastes of an elite. Moreover, the new class perspective began to establish new rules for the transformed genre: the action and content of the fairy tale subscribed to an ideology aimed at socializing young and old people according to the established values of the aristocratic class.

The example or lesson of *Beauty and the Beast* is an extreme one. Not all writing of fairy tales was as one-dimensional and class-biased. Nonetheless, once the folk tale began to be interpreted and transmitted through literary texts its original ideology and narrative perspective were diminished, lost, or replaced. Its audience was abandoned. As text, the fairy tale did not encourage live interaction and performance, but rather passivity. The perspective became that of the individual author, and there was a switch to either an aristocratic or a bourgeois point of view. The taste of the upper class audience and the control it exerted over publishing influenced the narrative perspective as well; thus, the social experience of all classes and groups of people was becoming increasingly mediated through the socialization process and technological changes in production and distribution.

The rise of the fairy tale in the Western world as the mass-mediated cultural form of the folk tale coincided with the decline of feudalism and the formation of the bourgeois public sphere. Therefore, it quickly lost its function of affirming absolutist ideology and underwent a curious development at the end of the eighteenth century and throughout the nineteenth century. On the one hand, the dominant, conservative bourgeois groups began to consider the folk and fairy tales amoral because they did not uphold the virtues of order, discipline, industry, modesty, and cleanliness, and they were regarded as particularly harmful for children because their imaginative components might suggest ways to rebel against authoritarian and patriarchal rule in the family. Thus folk and fairy tales were opposed by the majority of the middle class, who preferred didactic tales, homilies, family romances, and the like. On the other hand, within the bourgeoisie itself there were progressive writers, an avant garde, who developed the fairy tale as a form of protest against the vulgar utilitarian ideas of the Enlightenment. If we recall

Horkheimer and Adorno in *Dialectic of Enlightenment*, we can see that the struggle against what they called the instrumentalization of reason had great significance for the rise of the innovative fairy tale of the romantics, particularly in Germany. Even in the United States, the lines of opposition in the ranks of the bourgeoisie regarding the instrumentalization of both reason *and* fantasy are revealed in attitudes toward the fairy tale and imagination. Hawthorne ranted against the female writers of moralizing fairy tales, and Poe sought to frighten rationalistic bourgeois audiences with tales that chilled the Victorian sensibility. In England the battle over the moral worth of the fairy tale was especially fierce. As Michael C. Kotzin has pointed out in his book *Dickens and the Fairy Tale:*

> The cause for which the Romantics spoke came to have greater urgency as the conditions which provoked them to defend the fairy tale intensified during the Victorian period. Earnest, artless, middle-class Evangelicism increased its influence; the educational theories of the Enlightenment were succeeded by those of its even less imaginative descendant, utilitarianism; and the age of the city, industrialism, and science came fully into being. These conditions in England were objected to by Caryle and by such followers and admirers of his as Ruskin and Kingsley. In discussing the fairy tale these men followed the Romantics by stressing its imaginative value in the new world. But they also reverted a bit to the position of the enemy: the educational values they pointed to in the tales, while not usually as simply and exclusively instructional as those the Enlightenment advocated, are more conventionally moral than those which had been defended by Wordsworth and Coleridge. With their statements in defense of the fairy tale . . . the Victorian men of letters probably contributed to its new status. In those statements and elsewhere, they reveal the synthesis of appreciation of the imagination and moral posture which characterizes the Victorian acceptance of the fairy tale.[22]

Kotzin's remarks on the historical development of the fairy tale in nineteenth-century England are significant because they outline how the bourgeois public sphere gradually accomodated and instrumentalized fantastic art production to compensate for some of the ill effects of capitalist regulation. While it was the fairy tale's critique of utilitarianism that inspired the opposition of the Enlightenment, the established capitalism of the latter part of the nineteenth century could afford to be more indulgent. Watered-down fairy tales which were either moralistic in tone or served a compensatory cultural function, whose dissemination was facilitated by printing, became a useful means of alleviating the

---

[22]Michael C. Kotzin, *Dickens and the Fairy Tale* (Bowling Green: Univ. of Bowling Green Press, 1972), p. 26. The situation in England was similar to that in Germany. See Dieter Richter, Johannes Merkel, *Märchen, Phantasie und soziales Lernen* (Berlin: Basis Verlag, 1974), pp. 58–103.

tensions resulting from the increased regulation of daily life and its attendant alienation.

This is not to argue that the fairy tale was totally absorbed by the growing capitalist culture industry, as the following outline of the more significant tendencies in the nineteenth-century treatment of the genre makes clear: 1) After the Grimm brothers made their first collection in 1812, folk tales were gathered, transcribed, and printed for the purpose of establishing authentic folk tales. This was usually done by trained professionals who often stylized the tales, changed them, or were highly selective. Once gathered, the tales were rarely read and circulated among the original audiences. 2) Folk tales were rewritten and made into didactic fairy tales for children so that they would not be harmed by the violence, crudity, and fantastic exaggeration of the originals. 3) Folk tales were transformed into trivial tales, and new fairy tales were composed to amuse audiences and make money. Fairy-tale plays became fashionable, especially fantasy plays for children.[23] 4) Serious artists created new fairy tales from folk motifs and narrative structures. They sought to use fantasy as a means of criticizing social conditions and expressing the need to develop alternative models of social orders. 5) As new technological means of the mass media were invented, they incorporated the fairy tale as a cultural product to foster the growth of commercial entertainment or to explore the manifold ways in which fantasy could enhance the technology of communication and how the effects of fantasy could be heightened through technology.

All the above tendencies persist in the mass-mediated culture of the twentieth century. The advent of key technological inventions such as photography (1839), telegraphy (1844), telephone (1876), phonography (1877), motion pictures (1891), radio (1906), television (1923), and sound motion pictures (1927), has enabled the mass media to use the fairy tale on the one hand to affirm the interests of the culture industry by curtailing active social interchange and making audiences into passive consumers; and on the other, to communicate and unify cultural products of fantasy necessary for developing a more humanistic society and for stimulating audiences to play an active role in determining their destiny.

The culture industry mediates and interprets the experience of the people according to its marketability, and the major accomplishment of the mass media in the twentieth century has been to make it appear (unlike publishing) that the voice and narrative perspective of folklore emanate from the people's own cultural expression and heritage. It was the radio, then movies, and ultimately TV which were able to draw

[23]See Melchior Schedler, *Kindertheater* (Frankfurt am Main: Suhrkamp Verlag, 1972), pp. 43–71.

together large groups of people as the original folktale narrators did and relate tales as though they were derived from the point of view of the people themselves. Mass-mediated fairy tales have a technologically produced universal voice and image which impose themselves on the imagination of passive audiences. Whereas the original folk tale was cultivated by both narrator *and* audience to clarify and interpret phenomena in a way that would strengthen meaningful social bonds, the total meaning conferred by the narrative perspective of a mass-mediated fairy tale on reality has assumed totalitarian proportions, because the narrative voice is not only no longer responsive to an active audience but manipulates it according to the vested interests of the state and private industry. Nor is this the result of a sinister plot on their part; but rather of "unquestioned but fundamental socioeconomic arrangements that first determine, and then are reinforced by, property ownership, division of labor, sex roles, the organization of production, and the distribution of income."[24] As a consequence, the inevitable outcome of most mass-mediated fairy tales is a happy reaffirmation of the system which produces them.

III

In what ways has the fairy tale as a mass-mediated cultural form been used to further the interests of the culture industry? Have the utopian impulse and the alternative social orders of serious fairy tales been totally eliminated? Is there no room left in mass-mediated culture for the imagination? Has the bourgeois public sphere become so corrupt and powerful in its manipulation of public opinion that the creation of an alternative public sphere is no longer possible? A discussion of the fairy tale and its function in the mass media can help us assess the limitations placed on the imagination and counter-cultural forms and also point the way toward possible solutions.

The reception of fairy tales and folk tales in the Western world, and perhaps beyond, has been heavily influenced by preconceived notions of the fairy tale promoted by the Walt Disney industry and similar corporations. To counter this corporate inundation of our imagination, the familiar fairy tales must be made strange to us again if we are to respond to the unique images of our own imagination and the possible utopian elements they may contain. Otherwise the programmed fairy tale images will continue to warp our sensibilities through TV advertisements in which women are transformed into Cinderellas by buying new dresses and using the proper beautifying cosmetics, while the beast is turned into a magnificent prince by shaving with the right brand of shaving cream.

[24]See Herbert Schiller, *The Mind Managers* (Boston: Seabury Press, 1973), p. 5.

This widespread use of fairy tale and folklore material in the mass media has only recently been studied by critics. Raphael Patai, in his *Myth and Modern Man*,[25] is a case in point, although he is too eclectic in his examples and often confuses myth and folklore. More systematic are two studies on folklore in the mass media by Tom Burns and Priscilla Denby.[26] In a day's TV viewing in 1969, Burns recorded 101 traditional folklore items or themes and discovered that "four out of five advertisements using Märchen material draw upon familiar characters and plots and fit them into a product."[27] Denby's study went beyond that of Burns in that she collected folklore material from other media and expanded her categories to include folklore qua folklore (articles about folklore; folklore as foundation, mimetic folklore); folklore as folklure (for selling and decorative purposes); folklore as an aside. She reached the conclusion that folklore as employed in the mass media has nothing to do with its original purpose of fostering and reinforcing a traditional image, unifying people, and passing on tradition. It has become more international and universal:

> In a sense, the folklore as found in the media is meant to be a psychological magic wand; it can systematically inform and directly instruct one to revert to a simplistic life style, or it can subtly suggest that the world is somehow based on unreal forces over which one has no control anyway. It can thus serve as an escape, conscious or otherwise. Then again, it may also serve to demonstrate that society has become too technological to ever become "folk" in nature again, and, ironically, that the current emphasis on folklore and superstition which are sought as aids, only serves to further complicate and obscure the issues (a point of view prevalent in the more sophisticated or intellectual magazine). Thus, media folklore can be seen as a panacea or an obfuscation.[28]

Denby has placed her finger on how mediated folklore, whether as panacea or obfuscation, is instrumentalized to mislead audiences and mystify the cultural system. The system of mass communication and its links to capitalist commodity production are ultimately affirmed by folklorist plots and images which leave the audience happily passive or disturbedly impotent. Both Burns and Denby conducted their studies under the correct assumption that traditional folklore loses its strong cultural meaning when lifted out of its usual context of performance, setting, and audience. But, unfortunately, they do not discuss in depth how the new commercial context of mass media determines the function

[25]Raphael Patai, *Myth and Modern Man* (Englewood Cliffs, N.J.: Prentice-Hall, Inc., 1972).
[26]Tom Burns, "Folklore in the Mass Media: Television," *Folklore Forum*, 2 (1969), 99–106. Pricilla Denby, "Folklore in the Mass Media," *Folklore Forum*, 4 (1971), 113–23.
[27]Burns, pp. 98–99.
[28]Denby, p. 121.

and meaning of folklore, which, once it is mediated, loses its folk aspect. In the case of the folk tale, its mediated form is the fairy tale and its context is the culture industry.

Although the instances of this mediation are legion, I will limit my remarks to the case of the fairy tale as film, and even here, the analysis cannot pretend to be complete. It is intended rather to raise questions about the instrumentalization of fantasy, and since some of the generalizations about the fairy tale and film might seem exaggerated, it will be important not to lose sight of their socio-economic context—the Hollywood culture industry which judges its products largely according to standards of commercial success. Thus, no matter how culturally innovative, avant garde, or serious in intent the fairy tale as film may be, it will generally not be promoted if it does not meet the commercial and normative standards of Hollywood or of the TV industry which has developed its own market for films. Any analysis of the fairy tale as film must consequently begin with a critical premise regarding the commercial exploitation of both the gifted artists who produce films and the mass audiences who seek satisfaction from cultural products. The messages of the fairy tales however are not all completely rigged to trigger off the consumer mentality of the mass audience. As in the case of the literary fairy tale, where writers produced serious fairy tales critical of the market itself, and of the public sphere, within the commercial framework, so too with film. Therefore, after discussing two films which rely heavily on the fairy tale while affirming a capitalist ideology, I will explore their anti-capitalist and utopian elements. The films are *Snow White and the Seven Dwarfs* and *Rocky*, and the analysis will be confined to the narrative structure and motifs of folk and fairy tales used in them to sell the American dream of a free and democratic society. Hypnotic in form and content, they carry us off to a never-never land by means of remarkable and dazzling technical tricks, making us forget that we have out own unfulfilled dreams which are more important to project and fulfill than those that the film imposes on our imagination.

Each film uses the fairy tale differently. *Snow White and the Seven Dwarfs* is a direct animated adaptation of a Grimms fairy tale created for young audiences and families. *Rocky* is a sports film which is unconsciously based on fairy tales for a sports-minded general audience. Although there are other films which might be considered to be mass-mediated fairy tales, these two reveal some of the most interesting and dominant methods by which the fairy-tale structure and motifs have been appropriated by filmmakers to project their own fantasies about social conditions in capitalist society.

Snow White and the Seven Dwarfs, produced in 1936 as Walt Disney's first full-length animated feature, is most important because it became the prototype not only of all of Disney's other fairy tales and insipid family films, but for most feature film adaptations of folk and fairy tales. The circular pattern is dominant: an aging beautiful queen wants her step-daughter killed because Snow White is fairer than she. However, the young princess is saved by a royal hunter and eventually protected by seven dwarfs in the forest. The queen learns of this and transforms herself into an old crone who sells Snow White a poisoned apple. The dwarfs pursue the queen and cause her death. Snow White is apparently dead and encased in a glass coffin. But then Prince Charming arrives on the scene to give Snow White a kiss, and she is restored to life and power and rides off singing with her prince.

As Richard Schickel has noted, the numerous changes made by Disney radically altered the meaning of the Grimms' tale.[29] The tale involves a violent struggle over power with no holds barred, and it is told with lurid detail and powerful symbols which have deep psychological and social implications. The film neutralized all these elements and actually switched the focus to the dwarfs, who play a minor role in the tale. This refocusing in the mediation is significant, for it reveals the underlying ideological meaning of the film. The dwarfs are little workers, miners to be exact, and they sing and whistle while they work. Their names—Doc, Happy, Grumpy, Sneezy, Bashful, Sleepy, and Dopey—suggest the composite humors of a single individual. They represent the healthy instincts of a person. When he or she as viewer orders them properly, they can become a powerful force against the forces of evil. Maintaining order and the ordering of the dwarfs become the central themes of the film. Even Snow White, who becomes the dwarfs' surrogate mother, maintains order in their house. The images of the home and forest are all clean-cut, suggesting the trimmed lawns of suburban America and symmetrical living as models. To know your place and do your job dutifully are the categorical imperatives of the film. Snow White is the virginal housewife who sings a song about "some day my prince will come," for she needs a dashing male savior to order herself and become whole, and the boys are the breadwinners who need a straight mom to keep them happy. Though the wicked queen—the force of disorder—dies, the social order is not changed but rather conserved and restored with youthful figures who will keep the realm and their minds spic and span.

In order to understand this structured cleansing process in the film more clearly, it is important to bear in mind that America was still in the throes of a depression in 1936. Work was difficult to find, and workers'

---

[29]Richard Schickel, *The Disney Version* (New York: Simon and Schuster, 1968), p. 182.

discontent led to violent strikes and the rise of a strong socialist movement. Chaos, hunger, depravity were widespread. Government and industry sought all sorts of "new deals" to coerce and pacify the dejected people and to discourage active opposition, while maintaining the basic structure of the capitalist system. The image of America as one happy family pulling together to clean up the economic mess was important to disseminate if capitalism was to remain the dominant form of production in the society. Snow White as Miss America symbolizes the basic goodness of the American socio-economic system, and the dwarfs as workers order themselves neatly to defend this system. The need for security and a good home was basic to the dreams of the American people at this time, and Disney fed his audiences what they desired—but certainly not in their behalf. He reduced the wants and dreams of the American people to formulae: a prescription for gaining a measure of happiness by conforming to the standards of industry's work ethos and the constraining ideology of American conservatism. As Schickel has noted, *Snow White* signified the end of Disney's great experimental and eclectic period and the beginning of his instrumental phase—a phase which led him to build a corporate empire based on the instrumentalization of his art and fantasy to affirm and beautify the dehumanization process of the culture industry. Schickel asserts that

> eclecticism might have served as a brake on "Disneyfication," that shameless process by which everything the studio has touched, no matter how unique the vision of the original from which the studio worked, was reduced to limited terms Disney and his people could understand. Magic, mystery, individuality—most of all, individuality—were consistently destroyed when a literary work was passed through this machine that had been taught there was only one correct way to draw.[30]

The uniformity of the structure, style, and themes established in 1936 can be traced throughout his work. It was not, however, that Disney sought consciously to deceive his audience, but rather that his "popular" culture reflected the socio-economic conditions of his times, and hence resulted in a synthetic image of what America came to mean for him. That this image became destructive to the individual imagination and the emancipation movement of oppressed groups is a reflection of how Disney himself had become victimized and deluded by the demands of the culture industry.

Sylvester Stallone, the hero and writer of *Rocky*, says of himself: "To tell you a little bit about myself, I'm not that much more exceptional than any other actor. I've always maintained that maybe I had something

[30]Schickel, p. 190.

unusual, maybe I had something special that eventually I could sell, but the problem was finding someone who would buy that product—and that's exactly what it is: a product. If people think along heavily esoteric terms in which they are pure artists who won't ever sell out, they'll never make it because, unfortunately, the business revolves around the decimal point and the dollar sign; so you have to be artistic and commercial at the same time."[31] Thus, the underdog can only make it in America by succumbing to the market conditions of the culture industry. The film *Rocky* as a fairy-tale success story glamorizes and supports Stallone's statement. The structure of the film is based on the folk-tale motif of the swineherd who becomes a prince, and, to a certain extent, on the fairy-tale motif of the beauty and the beast. Here our lonely lower-class hero must show that he is of prince charming caliber in order to oppose the commercial interests of the boxing world. He is a brute or beast whose nature becomes more noble when he meets his homely and homey princess who is in turn transformed into a beauty once she grants the beast a kiss. It is the power of pure love plus the help of an old-time trainer and friends—the inevitable animal as friend is there, too—which enable Rocky to oppose the forces of evil symbolized by black heavy-weight champion Apollo Creed wearing the American colors during the Bicentennial in Philadelphia. Implicit in the film's structure, then, is a contradiction: Rocky sets out to prove the commercial shallowness of the American dream, and yet, he substantiates this myth by becoming the great white hope who successfully proves that the underdogs have something special about them and that the system does allow them to make it to the top. The ending also plays upon the racism of American audiences, for it is the white Rocky who represents goodness and integrity versus the black Apollo who represents evil and sham. Defeat is victory in this tale. Rocky is acknowledged the winner when Apollo whispers to him that he won't give him a rematch and when Rocky answers that he doesn't want one. Like most humble, modest Americans, what he wants most is Adrian, his pride, his love. Again the circular pattern of the fairy tale has a hero rise to the top and affirm the possibility of achieving happiness within the system.

There is a temptation to make Rocky into a noble working class hero and celebrate him for a lack of other more positive heroes in American films. As Ira Schor remarks, "watching him radiate in the seedy tenements, we want him to make it. He gets his chance."[32] But what is this chance in actuality? The opportunity to make a buck and buy into the respectable middle-class world in a seemingly dignified way. And why

---

[31] *The Official Rocky Scrapbook* (New York: Grosset and Dunlap, 1977), p. 9.
[32] Ira Schor, "Rocky: Two Faces of the American Dream," *Jump Cut*, 14 (1977), 1.

do we want Rocky to win? It is not so much because he is an anti-hero or counter-cultural hero, but because he apologizes for the so-called jaded and sordid conditions in America which we must accept if we want to survive. Rocky makes us feel a bit more comfortable with our lack of social conscience and the daily compromises we make to keep the system running smoothly. As a cheap hood, Rocky not only actively accepts crime, but he must learn to control his goodness lest it appear too neurotic. Stunted in his thinking and development, as is Adrian, he cannot achieve any sense of self-realization except through the boxing industry, which he knows is spurious. His acceptance of the hype— selling himself and the fight as product—even with his limited expectations and desire to retain his pride, is not an act of self-realization but rather a passive recognition and acceptance of the fact that he will remain a commodity for the rest of his life. It is exactly this condition which the culture industry induces us to accept. The illusion of a working class hero at the end of the film remains a deception because it does not speak to the working class impulses to abandon a condition which requires such an illusion.

But is *Rocky* really a film without hope for the subjective self-realization of human beings? Are the other fairy-tale films totally under the magic spell of the culture industry? In a discussion about the contradictions of utopian nostalgia with Adorno, Ernst Bloch stressed the positive side of those fairy tales containing fantastic elements of social utopia which resist the strictures of instrumental rationalization.[33] In contrast to the philosophers of the Frankfurt School, Bloch endeavored throughout his life to elaborate a philosophy based on the principle of hope. One of his primary interests was the exploration of popular culture, including daydreams, fairy tales, dress, advertising, decoration, and manners in order to locate their utopian potential for anticipating a better life on earth. In opposition to plain wishful thinking, Bloch maintained that there are concrete utopias which await their human fulfillment through the action of conscious individuals who learn through their conscious action how to take possession of these projections. These utopias have a political significance in that they do not affirm and strengthen the ruling classes in society but anticipate a new class and justify this emergence. Art works are significant in that they help induce what is not yet conscious (unfulfilled and repressed wishes, dreams, and needs) to become conscious so that human beings can seize upon these projections in order to realize their full potential. To Bloch's mind fairy tales, with their utopian elements, are crucial. The source of their

---

[33]Rainer Traub and Harald Wieser, eds., *Gespräche mit Ernst Bloch* (Frankfurt am Main: Suhrkamp Verlag, 1974), pp. 73–75, my translation.

attraction for the common people lies in their protest against repression:

> No matter how fantastic the fairy tale is, it is always clever in the manner by
> which it overcomes difficulties. Also, courage and slyness succeed in the fairy
> tale in a completely different way than in life, and not only that: Like Lenin
> says, it is always the existing revolutionary elements which pull together the
> given strands to spin a yarn. When the peasant was still bound by serfdom,
> the poor young hero of the fairy tale conquered the king's daughter. When
> sophisticated Christianity trembled upon encountering witches and devils,
> the fairy-tale soldier deceived witches and devils from beginning to end.
> (Only the fairy tale emphasizes how "dumb" the devil is.) The golden age is
> sought and mirrored where paradise was to be glimpsed in its depths. But the
> fairy tale does not let itself be imposed upon by the present owners of
> paradise. So, it is rebellious, a burned child and lucid. . . .[34]

Thus, according to Bloch, the masses respond unconsciously to
utopian signs out of need, and these are the signs which we must read,
interpret, and elaborate when we consider the great response of
audiences to such fairy-tale films as *Snow White and the Seven Dwarfs* and
Rocky. Though it is difficult to find a common utopian denominator
underlying these two films, their anti-authoritarian flavor is reminiscent
of the utopian, revolutionary component of the original folk tales which
was designed to bring people closer together so that they could make
their own history. In *Snow White and the Seven Dwarfs* the little people and
the frail Snow White unite to overthrow the oppressive rule of the evil
queen. In *Rocky* the downtrodden working class supports the underdog
in a system loaded to exploit the little man's efforts to gain dignity. In each
case the film as fairy tale projects the possibility of a better future with
freedom for the imagination if there is a united front against authoritarian
rule. In each case, the underdog (i.e., the youngest son or daughter)
serves as the symbol of the common people. As Bloch implies, there are
holes in authoritarian systems through which little people can crawl with
subversive results. But if this is to occur, the imagination must be kept
free to project the holes and alternatives to the dominant repressive
system.

Bloch claimed that his daily prayer was: "Grant me today my
illusion, my daily illusion. Illusions have become necessary today, vital
for life in a world that has been fully exposed by utopian conscience and
utopian presentiment."[35] It is its capacity to foster this necessary illusion
that Bloch admires in the fairy tale, but in doing so, he fails to take the
passivity of today's audiences and the control exerted by the culture

[34]Ernst Bloch, *Ästhetik des Vor-Scheins I*, ed. Gert Ueding (Frankfurt am Main: Suhrkamp
Verlag, 1974), pp. 73–75, my translation.
[35]Traub and Wieser, p. 73.

industry over the mass-production of aesthetic illusions into account. Whereas formerly the utopian aspect of the fairy tale (which also contained conservative elements) was preserved by the context in which it was actively received and retold, when it is transmitted today, as text, film, play, advertisement or TV show, the narrative voice is actually controlled by commercial interests. Despite all the possible utopian images contained in the narrative structures of the films which certainly contain anti-capitalist tendencies, they cannot have a liberating effect: first, because of the context in which they are embedded, the circular structure of cultural affirmation, and second, because of the context in which they are transmitted, received, and circulated, i.e., the culture industry. Bloch's theory can only lead us to pose the question of how we might liberate the utopian elements of popular culture, since the instrumentalization of reason and fantasy has not yet been completed.

The first step toward resisting the debilitating practice of the culture industry and guarding the imagination from instrumentalization must be to create a counter public sphere which could lend force and expression to groups opposed to the systematic alienation that results from commodity production. Indeed, a radical movement might exploit the very collective nature of the technological mass media. As Real has pointed out: "Crucial to the potential of the newer media is their *collective structure*. As opposed to the individualism of writing and reading, the new media call forth collective efforts of social organization."[36] The problem remains one of stimulating people to organize actively around their interests. In contrast to the pessimism of the older members of the Frankfurt School, Real argues that

> Mass-mediated culture can be rendered more understandable and humanistic by increasing cultural studies and policy research, by continuing consumer activism and reform efforts, by reducing the role of private capital and profit, by reversing one-way authoritarian transmission, by decentralizing, by developing in the wake of structural revolution a cultural revolution that returns to top priority the full collective humanity of persons, the value of life, and the appreciation and balanced development of the environment.[37]

Given the manner in which the fairy tale is locked into mass-mediated culture, its own structure and capacity to convey creative images of emancipation depend on what Real calls for: a radical reordering of the public sphere. The fairy tale may contain utopian elements, but it will not become history unless people learn how to depend on themselves. Here we return to Novalis, Benjamin, and Horkheimer, but the return is not the traditional circle, for all three

[36]Real, p. 267.
[37]Real, p. 268.

attempted to break with conservative and rational thinking to glean from the fairy tale what was lacking in actuality— the reappropriation of humankind's own productive and creative value by conscious human beings. This lack is still there, and the fairy tale, though mediated, still projects rays of hope that humankind will come into its own.

The Political Economy of
Social Space*

ANDREW FEENBERG

ll points in a geometric continuum are equal; social space could not be more different. Insofar as it is socially significant, distance is no mere natural fact, but is socially produced and socially functional, its significance relative to the dominant mode of production. For example, in capitalist societies and the world economic order they dominate, location has a strict class character: the distribution of people and goods in space corresponds to a hierarchical system of classification. At the same time, the "conquest" of space by new technologies of transport and communications contributes to the development of new forms of individuality, less provincial, more cosmopolitan, enriched by access to the wide experience of the species.

This paper will consider the implications of such changes in social space in the light of Marxist theory. Marxism has no explicit theory in this domain, but it does contribute to the understanding of the two major transformations of social space in modern times, globalization and urbanization. The spatial pertinence of the Marxist theory of globalization has been brought to the fore recently by dependency theory. Wallerstein and his collaborators explain the "world system" in terms of a hierarchy of geographical locations, to each of which is attached a different role in

*Research for this paper was supported in part by the California Council for the Humanities in Public Policy, and the Center for Twentieth Century Studies of the University of Wisconsin-Milwaukee.

the international division of labor. Here I will be primarily concerned with the application of Marxist theory of social space to the city. Engels argues that urban life is intellectually liberating for the lower classes. On the basis of Marxist theory, it can also be shown that the capitalist city distributes classes spatially through the social allocation of space as a commodity, a reward for socially validated performances.

To some extent globalization and urbanization have had the consequences foreseen by Marx and Engels, especially in terms of the rise of new forms of individuality. However, in this domain as in so many others, the twentieth century has witnessed the development of powerful counter-tendencies which Marx and Engels did not anticipate. Where the latter saw the capitalist reconstitution of urban space as essential to the rise of the socialist movement, both mass communications and the automobile have contributed to the reintegration of the underlying population around a highly atomizing urban experience. This urban world is entering a new crisis today, for which no solution has yet appeared.

I. The Synthesis of Social Space

Edward Hall calls social space the "hidden dimension" of human experience.[1] In what sense is space hidden? In fact nothing would seem to be more manifest than the space through which we move in everyday life. Yet Hall is right to suggest the invisibility of the specifically *social* space of daily life. This socially constructed and perceived space of ordinary experience is hidden to us by a naive objectivism that leads us to conceive distance as a natural fact which we alter occasionally for utilitarian purposes. On the contrary, social space is constituted objectivity, a "lived" reality and not a natural given. The study of such lived realities was initiated by phenomenology, which is worth considering, if only briefly, for its methodological implications.

Husserl founded phenomenology by "bracketing" what he called the "natural attitude," the everyday naturalistic assumptions made in the course of ordinary thought and action. He regarded the objectivistic concept of nature we usually take for granted as in fact the result of a complex "positing" or construction based on an immediate experience of the world prior to objectivity itself. Phenomenology was designed to expose the texture and organization of this immediate experience. Following Husserl, Heidegger later bracketed objective space, as a pre-existing quantitative continuum, and set out to explain it in terms of the experience of distance and movement of finite, spatially located subjects.

While phenomenology's discovery of such lived space represents a significant advance over objectivism, and helps us to focus on the

---

[1]See Edward Hall, *The Hidden Dimension* (New York: Doubleday, 1966).

"hidden" dimension of experience, its individualistic bent limits its usefulness for the study of social phenomena. Social space cannot be constructed from individual lived experience, for it shapes that very experience; the constitutive process in which a spatial reality such as a city takes shape is not individual but social in character. The production and function of distance are related to the project of society as a whole, the values in terms of which it constructs its world of social objects in the practice of social life. Hall and other anthropologists have shown that these values can best be interpreted through the use of the concept of "culture." Nevertheless, although it may be necessary to abandon certain aspects of phenomenological epistemology, the basic concepts of bracketing and lived experience can help to clarify the distinction between nature and culture so necessary to a cultural analysis of space.

There have been some interesting attempts in recent Marxist theory to define the boundaries between nature and culture. Jürgen Habermas has introduced the term "materialist synthesis" to explain how natural objectivities become social ones through their incorporation in a social-life process.[2] In this sense, presumably space too would be subject to a materialist synthesis. Habermas formulates his concept of materialist synthesis in Kantian terms, as the imposition of schemata of experience on a pre-existing nature. From this standpoint it can be shown that technology plays a basic role in the constitution of social space. Certainly, the social quality of distances depends on the means of traversing them and cannot be derived from their natural measure. Technology is thus not simply a means to a pre-given end, a phenomenon to be understood in strictly utilitarian terms as more or less efficient, but rather it too must be suspended in its quasi-naturalness and studied as an integral element of culture. Construed as such, technology is an artefact possessing a specific social meaning as well as a "use" in the ordinary sense of the term. Indeed, the application of technology is the process of inscription of this meaning on the natural presuppositions of social life.

Considered naturalistically, the technologies of transportation and communication through which space is manipulated appear to overcome a pre-existing immediacy—objective distance. But more importantly, these technologies must be understood as *positing* the distances they traverse. The automobile, for example, not only makes it possible to travel more efficiently through objective space, it also contributes to the construction of a new type of social space in which, among other things, the distances between things increase immensely. I will consider in more detail later the sense in which the automobile can be said to constitute the distances it traverses.

[2]For a discussion of this concept, see Jürgen Habermas, *Knowledge and Human Interests*, trans. J. Shapiro (Boston: Beacon, 1971), pp. 25–36.

II. Marxism and the Politics of Space

Although the writings of Marx and Engels do not address the problem of space in the encompassing terms introduced above, the transformation of social space does play a fundamental role in their theory. For Marx and Engels, capitalist society accomplishes the mediation, or socialization, of all "natural" presuppositions of life, thereby preparing the development of the higher forms of individuality and social organization associated in their theory with socialist society. For example, capitalism's universal process of exchange is supposed to diminish the influence of factors such as family origin, race, and sex in determining personal destiny. Similarly, the immediacy of spatial location is supposed to be mediated by the capitalist "conquest" of space and the concomitant breakdown of ties to the land, local customs, and authorities, with ultimately liberating consequences. The capitalist control of space is, however, "ambivalent," in Ferenc Feher's sense of the word: it promises and to some extent delivers a level of individuality it cannot completely fulfill and therefore points toward socialism as a resolution of its contradictions.[3]

Writing at the very beginning of the industrial era, Marx foresaw the planetary character of the civilization coming into being. Indeed, industrial society is unique in the history of social formations because it affects the whole globe simultaneously, gradually synchronizing all of world history and integrating the planet's most distant regions into a single continuum of interactions and interdependencies. This globalization of history, the emergence of world history as a positive reality, is due in great part to the development of new technologies of communications and transport which leave their mark on every facet of human life. As Marx writes, "The transformation of history into world history is . . . a completely material empirically verifiable act, an act for which every individual furnishes proof as he comes and goes, eats and drinks, and clothes himself."[4]

The transformation to which Marx refers involves more than simply bringing all the points of the globe into contact: more important is the contact so established between the independent cultural acquisitions of the various peoples of the earth. The synthesis of all these cultures through planetary interactions changes the relation of the individuals to each of them, and indeed to culture generally. Where formerly individuals necessarily conformed to the particular culture in which they

---

[3]For a discussion of the application of the concept of ambivalence to technology, see Andrew Feenberg, "Transition or Convergence? Communism and the Paradox of Development," in *Technology and Communist Culture*, ed. Frederic Fleron, Jr. (New York: Praeger, 1977).

[4]Marx, "The German Ideology," in *Writings of the Young Marx on Politics and Philosophy*, ed. Easton and Guddat (New York: Doubleday, 1967), p. 429.

were born and raised, now the effectiveness of the pregiven forms of
existence which each culture attempts to impose on its members is
weakened by contact and comparison with other cultures. Industrialism
is thus accompanied by a certain liberation of the individual from culture
as such. Marx writes:

> In fact, however, when the limited bourgeois form is stripped away, what is
> wealth other than the universality of individual needs, capacities, pleasures,
> productive forces, etc., created through universal exchange? The full
> development of human mastery over the forces of nature, those of so-called
> nature as well as of humanity's own nature? The absolute working-out of his
> creative potentialities, with no presupposition other than the previous
> historic development, which makes this totality of development, i.e. the
> development of all human powers as such the end in itself, not as measured
> on a *predetermined* yardstick? Where he does not reproduce himself in one
> specificity, but produces his totality? Strives not to remain something he has
> become, but is in the absolute movement of becoming?[5]

In this "absolute movement of becoming," mankind is the product of
global interactions that subvert pre-established ways of life, throwing
them all into the wild dance of universal equivalence. The technological
mastery of the forces of nature, including the conquest of space, does not
only serve utilitarian ends, but also brings about a change in the essence
of what it is to be human. Marx discusses this change only in negative
terms, as the overthrow of limitations. While the outcome of this process
of overthrowing limitations cannot be foreseen, the process itself can be
explained in some detail.

It is in the nature of a finite being that its location in space is a part of
its being. As I have argued above, for a human being, location entails
specific cultural as well as physical constraints. From Herodotus on, the
world traveler perceives all cultures as contingent because they are all
bound to a specific location. Until the industrial era, however, this ability
to transcend the culture of one's birth is rare indeed; almost no one has
the means to challenge the claim each culture makes to be universal,
defining what it is to be human. Limited in its mastery of space in the pre-
industrial era, the human species thus lives almost entirely under the
power of local gods.

Marx argues that with industrialism, however, humankind is cut
loose from these ties to various places and cultures. There is a Prome-
thean implication to this thesis: the local gods fall as industrial man
"storms the heavens." Yet the passage from Marx quoted above shows
that his Prometheanism implies no desire to make a god of man; he did
not believe the human race would achieve a fixed perfection outside time
once having overthrown local limitations. Marx aims at release from the

[5]Marx, *Grundrisse* (Baltimore: Penguin, 1973), p. 488.

fixity of local gods to unrestricted self-change. In sum, the full realization of human finitude as endless flux and change requires the "conquest" of space as its technological precondition. Only on this basis can "Spirit," in Hegel's words, be "externalized and emptied into Time."[6]

Marx develops this vision against the background of a critique of the capitalist disposition of global space. He regards the capitalist creation of world history as a great positive achievement, while condemning its specific forms, such as imperialism. Marx observes that "separate individuals, with the broadening of their activity into world-historical activity, have become more and more enslaved to a power alien to them. . . ."[7] This power is, generally speaking, the power of capital, the combined and accumulated means of production created by industrial society. In one of its aspects, however, it is space itself which strikes back at those who dare to "conquer" it. The unintended consequences of so many unprecedented, unplanned interactions overwhelm the individuals who initiate them. Liberated from space, the individuals are no longer protected by distance from their enemies and from fortune.

The "revenge" of abused space is documented in Marx's work under several headings: the constant warfare and social dislocation accompanying more frequent and intense interactions between nations and peoples; the reorganization of global space around an imperial hierarchy of wealth and power, at the expense of non-Western cultures and lives; and the unintended consequences of urbanization, such as the problems of pollution and disease in the cities and the impoverishment of soil in the country. The last of these is the most important for Marxist theory, because it is in the crucible of the city that the agents of liberation are supposed to be formed.

For Marx and Engels, the city is more than just a means for organizing space in the interests of industrial production; it is also the creator of original values. The city generates a human type which would not have existed without the transformation of social space the city itself brings about. In the city, for example, people learn to desire things that they might not even have been able to imagine before. As a mechanism for the control of space, the city has a special role in enriching personality and individuality; it multiplies human complexity by increasing the range of choices and the number of interactions people can engage in. As Marx puts it, " The real intellectual wealth of the individual depends entirely on the wealth of his real connections."[8] The city is not only instrumental in increasing the wealth of real connections, it is co-extensive with it.

---

[6]Hegel, The Phenomenology of Mind (London: Allen and Unwin, 1961), p. 807.
[7]Marx, "The German Ideology," p. 429.
[8]Marx, "The German Ideology," p. 429.

Through its control of space, the city becomes both the producer and the product of richer, more fully developed individuals than can exist in a rural society, poor in the control of space.

This enrichment of individuality is an unintended consequence of the capitalist control of space which promises an end to capitalism. For Marx and Engels, mobility in space is a defining characteristic of the proletariat and a secret ally in the class struggle against capitalism. Engels writes that "for workers in the big cities freedom of movement is the prime condition of existence. . . ."[9] He links this freedom of movement to the breakdown of authoritarian consciousness, a necessary pre-condition of socialist revolution:

> The hand weaver who had his little house, garden and field along with his loom was a quiet, contented man, "godly and honourable" despite all misery and despite all political pressure; he doffed his cap to the rich, to the priest and to the officials of the state and inwardly was altogether a slave. It is precisely modern large-scale industry which has turned the worker, formerly chained to the land, into a completely propertyless proletarian, liberated from all traditional fetters, *a free outlaw;* it is precisely this economic revolution which has created the sole conditions under which the exploitation of the working class in its final form, in capitalist production, can be overthrown. And now comes this tearful Proudhonist and bewails the driving of the workers from hearth and home as though it were a great retrogression instead of being the very first condition of their intellectual emancipation.[10]

Engels understood, of course, that anti-authoritarian consciousness would be insufficient to effect real social change by itself, without the emergence of new forms of solidarity to replace the old ones of the repressive rural *Gemeinschaft* destroyed by capitalism. This was to be the contribution of unions, parties, and cooperatives in the city. But there is an aspect of this process of re-association insufficiently clarified by Marx and Engels.

Urban social organization is "abstract," impersonal, as compared with the concreteness and substantiality of rural relations. Where people are born in neighboring houses and grow up and live side by side, the unmediated fact of their spatial proximity becomes a solid basis for relationship. But highly mobile urban workers must constantly enter into new relations with new individuals, their bonds formed on the abstract basis of common class experiences and interests rather than on the concreteness of nearness. Just how likely is this substitution of class for spatial commonality as a basis for solidarity? Are not privatization and atomization at least as likely a consequence of the mobility of the

[9]Engels, *The Housing Question* (Moscow: Progress, 1970), p. 45.
[10]Engels, p. 29.

individuals? An examination of the contemporary city shows to just what extent the capitalist appropriation of social space has succeeded in obstructing the revolutionary possibilities in which Marx and Engels were confident precisely by weakening most bonds of solidarity.

III. The Distribution of Social Space

It is significant that Marx and Engels write about urban life without any nostalgia for earlier communitarian phases in social development. For them the city is the place of increasing freedom and of a public life accessible to the lower classes, even if under capitalism it is the place of ruthless exploitation, disease, and demoralization as well. Yet, we who live in the closing decades of the twentieth century may well look back with a certain nostalgia at the cities Marx and Engels lived in and described. At least in one important respect—their role as the geographic locus of the public sphere—those old cities were greatly superior to our own. The difference is largely to be accounted for by reference to the automobile.

To understand the impact of the automobile on urban life, it is essential to focus on its cultural significance, particularly its effects on the public sphere. Traditionally, the city has been organized around a privileged space of public encounter, great palaces and squares, gardens and promenades which, at least after the bourgeois revolutions, were widely accessible and invested with deep symbolic significance as the spaces corresponding to the status and role of citizenship. In the modern city, this system of organizing space for *public* purposes has been shattered by the automobile and replaced by a new private system of distributing transportation opportunities. As we know, in contemporary capitalism, control of space is distributed privately—cars, plane tickets, and so on—and indeed the personal control of space has become one of the most important rewards offered by society. The organization of public space no longer plays an important role, subordinated as it is today to the problems of distributing control of space as a commodity.

This reconstitution of social space required a new technology—the automobile—but it cannot be understood as a simple response to that technology. Rather, the new social space responds to imperatives of the capitalist system which determine the functional role of the automobile in the society. There imperatives are, quite simply, the unending search for ways of extending the reach of the commodity form. Capitalism depends on the motivation of individual enterprise through the use of a highly articulated system of rewards, distributed individually on the market in direct proportion to personal performance. The transformation of a good into a commodity is the essential step in its integration into the system of rewards.

But not all important goods can be distributed individually on the market at any given level of technological and social development, nor is it always profitable or politically acceptable to produce goods people need. Even under capitalism non-economic channels, such as politics, voluntary common action, or personal activities outside the market, arise as people pursue values capitalist enterprise cannot supply.[11] Traditionally, urban space was such a public good, not distributed individually as a reward for performance, but freely available to all members of society. Today, the city as public good has contracted to its vestigial redoubt, the public park which cannot be sold as a commodity even in the age of the automobile.

The privatization of urban space is similar to many other transformations that have occurred in the history of capitalism. Beginning with the village commons, public goods are increasingly distributed as private ones because in their public form they do not motivate economic activity. Indeed, the more goods are publicly distributed, the narrower the range of goods available to motivate socially validated performance. Public goods are in fact goods distributed according to need, in something like the Marxist sense: everyone has a right to use such goods insofar as they are human, apart from their putative contribution to society through labor or investment.

Many of the irrationalities of the capitalist distribution system are due to the attempt to distribute goods inappropriate by their social nature to private distribution, such as contributions to knowledge (which may be allocated through patents, for example), education, health care, and, as I will show, social space. Viewed from the standpoint of a social accounting, the private distribution of control over space is intrinsically irrational and involves an enormous amount of effort for a relatively small result.[12]

Economists usually measure the value of goods by what people are willing to pay for them. That is a useful indication to a buyer or a seller, but it often gives no insight into the value of complex goods such as control of space. To study such a good it is necessary to use a different method of social accounting. As a first approximation to a more adequate accounting method, I suggest that the whole organization of social space and transportation be evaluated in terms of its ability to meet human needs and to develop complexity of individuality through expanded

[11]For a discussion of these matters, see William Leiss, *The Limits of Satisfaction* (Toronto: Univ. of Toronto, 1976).

[12]For another approach to this conclusion, see K. William Kapp, *The Social Costs of Private Enterprise* (New York: Schocken, 1971), pp. 197 and following. Kapp's argument is based on the problem of waste, due to competition and imperfect regulation. These are clearly important concerns, from which I abstract here.

interaction. Many uses of transport in capitalist society do not meet this criterion, even though they may have a high value as measured by cost. Two major types of irrationality account for this gap between the contribution of capitalist transport to human welfare and its high cost.

1. *The Dialectic of Centralization and Decentralization.* Improved transport is often said to give rise to decentralization. In reality, this is only half the story, for to the decentralization of some facilities corresponds the centralization of others according to new patterns. In the United States, residences are typically decentralized while services and places of employment cluster in multiple centers. The more efficient the transport system, the wider the area that can be reorganized around this new pattern. The implication of this dialectic is a paradoxical one: the availability of transport tends to increase the distance between things that used to be close together. In this way transport creates its own demand.

Cities discover this dialectical feature of transport in dealing with congestion. They may build new roads and more public transport, but the effect of this is to increase the property values of land on the perimeter of the city. That land then becomes too valuable for farming and is sold for suburban development. Soon there are still more people trying to get in and out of the city and the problem is posed all over again. This is why some transportation experts regard congestion not as a problem, but as a normal consequence of combining planned transport with unplanned land use. Under these conditions, once transport is available residences decentralize, services are further concentrated, and the distances between things increase inexorably.

It is often argued that centralization of services increases welfare through economies of scale. This is true is many cases, but often all that can be shown is that it is more profitable to operate the centralized facility, or that centralization allows the control of certain costs, but not that the increased profitability or the control of costs contribute to the overall welfare of consumers. I cannot argue this general point here, but in some cases it is patently clear that welfare is not increased by the ability to traverse the distances transport itself creates.

This is especially obvious in the distribution of consumer goods. It used to be that on every block there were small shops where consumers purchased goods from familiar people. Now one drives a car to and from the shopping center. That does not seem to increase anyone's welfare except the owners of the chain stores, although it is true that the money spent on it is counted into the usual measure of national income, the GNP. In fact, all this "progress" seems to leave consumers where they were at the start. They obtain the same goods: the only difference is that now they have driven a car to a centralized location instead of walking to a few shops in the neighborhood where they might pay a bit more, but would save on gas, insurance, taxes, and so on. The effect of transport in this instance is to create distances and then to cross them, with a net increase in welfare of zero.[13]

2. *The Problem-Distributing Function of Transport.* Were these paradoxical consequences of capitalist transport the only irrationality in the system, it might be relatively easy to mobilize the demand for change. However, there is a second socially integrative aspect of transport that does not add anything to social welfare, although it does add something quite significant to many individuals' welfare. It is this second irrationality which binds people tightly to the system. The problem-distributing function of capitalist transport consists in the mechanisms that allow some people to escape problems that are created by (or perceived as being created by) other people or institutions. For example, in a city where there are polluting factories or a great deal of crime, people will naturally seek to escape these "discommodities." Those who are wealthy enough to escape will do so by buying a house in the suburbs and driving to work, polluting on their way the very regions they do not wish to live in.

Problems such as pollution are, of course, social in character and ought to be solved by society equally for all. In fact these problems are not solved, but rather become occasions for the normal pursuit of economic rewards; they are integrated into the society's system of incentives for individual achievement. This is accomplished by distributing people in space at a greater or lesser remove from the problems. Those distributed close to the problem are the poor, whose activities are not rewarded highly by society, while those distributed further from the problem are people whose activities are most highly rewarded. The social problem does not get solved as such, but is instead integrated into the system of incentives provided by the society for motivating individual achievement. Thus in great part we can explain the specific spatiality of the American city by studying the distribution of space as a representative or map of the class system.

These examples show that transport makes a much smaller contribution to social welfare than it might seem from the amount of money (and time) individuals spend on it. Many of the functions of transport accomplish nothing positive, dealing either with the problems transport itself creates—the distances—or palliating social ills for some individuals who can afford to get out of the behavioral sinks of society. No system of private distribution of social space can solve these problems, for within

[13]Barry Commoner has made a similar analysis of the effects on welfare of a number of other economic changes that amount to little more than new ways of delivering familiar goods, although they are constantly touted as great advances. See "Dispute: The Closing Circle," *Environment*, 14, No. 3 (April 1972), 23 and following.

the terms of the system itself these are not really problems at all. In order for transport to make a larger contribution to enriching human life, it would have to be treated as an element in a broadly conceived social distribution of space for social purposes.

IV. The Decline of the Public Sphere

Americans spend approximately $120 billion a year for transport, eighty-six percent being spent on cars, which are used mostly for shopping and commuting. I have already discussed shopping above; the same principles apply to commuting. People do not wish to live near the ugly, noisy, smoky workplaces capitalism builds: they would rather drive great distances simply to avoid proximity to such devastated environments as those which house American industry. But one can imagine a different way of building and operating workplaces, compatible with a different distribution of the individuals in relation to them. In fact, different ways of distributing goods existed just a short time ago in all American cities. A rational transport policy would attempt to reconstruct social space to reduce the need to use cars for shopping and commuting. The absolute use of transport might then decline, while its contribution to human welfare rose.

What I am suggesting is the need to socialize urban space, to make the control of space a public matter through a political process in order that space itself may become once again a public good. It is obvious that this is not happening now. An atomized population finds it difficult to provide itself public goods. To the degree that this atomization is due to the capitalist appropriation of social space as a private good, to that degree the problem is not only a problem, but also raises cultural barriers to its own solution. As with many other problems in advanced capitalist society, this one appears as an occasion for change only abstractly, in theory; in practice the cultural impact of the problem masks it, de-mobilizing potential opposition. Thus, no longer providing motives for change, the problem may persist and even intensify for an extended period.

In an earlier period, when Engels, for example, was writing about these matters, workers were concentrated in urban centers invested by tradition with great public, symbolic significance. Workers could mobilize rapidly to seize these centers in great political actions, and the habit of mobilization was itself acquired in frequent labor struggles. The dispersed workers of contemporary American society, scattered through endless "bedroom communities," organized by institutionalized unions that discourage mobilization in any form, find it much more difficult to imagine and engage in political action. To be sure, greater mobility has made it possible for these workers to achieve a higher level of personal dignity and independence than could the rural workers Engels so despised. But in destroying traditional "natural" communities, greater mobility has isolated individuals, making it difficult to combine voluntarily for any end. Elementary organizational abilities and dispositions have been lost as basic non-political social units disintegrate. What is more, motivations to re-establish new voluntary organizations are weakened by rapid movement of individuals who must come together to create them; after all, it is difficult to combine to produce a common good that will be appropriated collectively over a long period when one expects soon to be on the move, unable to benefit from an investment made in the community left behind.

Thus, to an ever-increasing extent, the only important investments made by the great majority of individuals are in their personal, private fragment of the infinite continuum—the house and car in which is tied up their whole capital of wealth and status and which represent more than anything else their place in society. Even public goods and services are subordinated as much as possible to this privately distributed bit of social space. The buyer of a house, for example, calculates into his estimate of its worth the value of the associated school system, the quality of the air, and so on. Under these conditions, economic perceptions evolve in a way extraordinarily unfavorable to the development of political consciousness and action. Individuals come to see goods exclusively in terms of what can be distributed on the market for private appropriation. Public goods are increasingly regarded as expenses without returns, as "waste," if they cannot be acquired in some obvious form indirectly through a market mechanism. These perceptions replace the traditional ones that used to bring people into the political process in an attempt to enhance their welfare through it.

The right to engage in the politics of space is formally recognized by the bourgeois state, which establishes agencies for the ostensible public regulation of such matters as zoning and transport policy. The irony is that the multiplication of boards, regulations, public hearings occurs in a society so atomized and privatized by the capitalist distribution of space that individuals hardly make any use of their formal rights. From the standpoint of many defenders of the present system, the decline of the public sphere, to which this indifference testifies, is highly desirable. It makes it possible to control the rise in public expenditures for public goods, thereby chaining the population that much more tightly to the system's reward structure. What is more, only "responsible" elements are then heard from in the political process. Under these conditions, openness—the formal right of participation, public hearings, and so on— does not serve democratic purposes, but those of business instead.[14]

[14]This is the argument of many observers, for example, of Sheldon Novick in *The Electric War: The Fight over Nuclear Power* (San Francisco: Sierra Club, 1976).

Business alone is organized and clearly motivated to take advantage of its "democratic" rights. Tremendous distortions in the structure of social space ensue. It is these distortions, in turn, which sustain and reproduce a privatized world in which the public sphere has no locus.

The implications of all this go beyond the costs of economic irrationalities, great as those may be. Behind the utilitarian concept of transportation as a means to its users' ends, there lies a deeper connection between transportation and human individuality. The purposes for which transport is used reflect the values of the users. In serving these values, the transport system and the social space it organizes become an outer spatial expression of the human needs recognized by society. If we add all the uses of transportation together, the sum is a whole person, whose different dimensions can be mapped on the city's surface. Social space is thus a macrocosmic representation of the individual or, rather, of what society determines the individual to be: it is an objectification of the socially given definition of individuality. To give the organization of social space over wholly to business is thus to subordinate all dimensions of human being to those primarily serviced by business, especially those compatible with its free, unfettered development.

What can we say of the results? Who would have thought they would be so catastrophic only a few decades ago?

In Plato's *Republic*, the city serves as a metaphor for the individual soul. The perfection and disorders of the one can be traced in the other. The analogy is perhaps more compelling today: it can be concluded that the condition of the American city testifies to a chaos and despair in the individuals beyond Plato's wildest imaginings.

Irony and Anarchy:
Technology and the Utopian Sensibility

DAVID L. HALL

The primary function of utopian theory is to critique the assumptions of our classical social and political theory, thus providing speculative alternatives which will help us understand our social world. But since, as I will argue, the Anglo-European tradition of utopian speculation is rooted in the same fundamental theory of origins as all our classical social and political theories, and since contemporary technological society requires a new vision, our utopian speculation is not radical enough. Utopian speculation must deny the traditional Western belief that social order must be grounded in principles which are external sources of that order; it must, in other words, be radically an-archic. I will turn to Taoism for a model of anarchism relevant to technological society, suggesting that one of the possible positive effects of our technological society is that it allows us to turn away from classical politics and toward social activity which can promote anarchism.

I. Utopia and Anarchism

In Western culture, the utopist accepts the same problematic as classical social and political theorists: how to erect a viable society on a foundation of rational and moral principles. This presupposes a deep-rooted philosophical conception of order, which makes its first appearance in the cosmogonic myths we have constructed to tell the story of our origins.

These speculations on the origin of the cosmos have generated the fundamental theoretical structure by which we think about "the World."

Three principal accounts of the origin of things are to be found in the mythical resources of Anglo-European culture. In *Genesis*, influenced by the Babylonian *Enuma elish*, an act of Divine volition provides the foundation of rational and moral order (depending upon one's interpretation, creation in *Genesis* is either *creatio ex nihilo* or it is the construal of order from the primal confusion of "the dark, formless, void"). In Plato's *Timaeus*, the order of the world is the result of rationalization—the conquest of blind Necessity by the rational persuasion of the Demiurge. In Hesiod's *Theogony*, it is neither volition nor reason, but the attractive power of Eros which overcomes the "yawning gap," the "gaping void," of Chaos which resulted from the separation of Earth and Sky.

In all these accounts, Chaos is characterized as a dark emptiness that sets us to brooding or leads us to despair; it is a disordered confusion confounding our sensibilities, a chasm consuming us with yearning. As acting, thinking, and feeling creatures, our fundamental cosmological problem is the maintenance of order in the face of such Chaos. We are literally *agonal* creatures whose contest with Chaos is the defining characteristic of our natures.

It is a commonplace among scholars of ancient Greek culture that the word *kosmos* (κοσμος) was first used only in terms of artificial or humanly constructed forms of order or arrangement. Thus, the verb *kosmeo* (κοσμεω) and the plural *kosmoi* (κοσμοι), for which there are no viable English equivalents, were commonplaces in the Greek vernacular. Until the appearance of the earliest natural philosophers, the *physiologoi*, there was no generally accepted notion of the cosmos as a single-ordered, systematic whole. There were *kosmoi*, but there was no *Kosmos*. The transition in cultural consciousness from the conception of a relatively haphazard Pluriverse, in which supra-natural gods intermittently intervened, to the notion of a single-ordered Cosmos, a Universe bound by rational and moral laws, occurred *pari passu* with the theoretical development of Greek intellectual culture. And this theoretical development is the story of a series of "cosmetic" activities designed to construe order from out of Chaos.

The *physiologoi*, our first cosmologists, sought principles (*archai*) which would establish the basis for the order of the world. These principles, functioning as external determining sources of order, were means of disciplining Chaos. Thus, from the very beginning, our theories were instruments for construing the world. This may easily be seen in the social theory of Plato's *Republic*, a work that has significantly influenced both classical and utopian speculation in the West. Plato sought principles of functional specialization and education which were to be

Universals, Eternal Forms, norms for thought, action, and passion. The Principle of the Good, the ground of all principles of normative measure, was at once the presupposition of dialectical enquiry (Platonic pedagogy) and the final goal of that enquiry. The principles of specialization were grounded in the analogical relations between the functions of the psyche (thinking, acting, and feeling) and the classes of the model society (the philosophic rulers, the guardians, and the craftsman, or technical class). The primary virtues (wisdom, courage, and temperance), which promoted justice in the soul and the state, were both the means and the end of functional specialization.[1]

Thus, in the Western tradition, thinking, acting, and feeling—the primary means by which we recognize our humanness and the humanness of our world—are instruments for overcoming Chaos. The well-nigh total acceptance of reasoning and moral praxis as the means of establishing social order, is thus tantamount to the affirmation of the primacy of disorder, of negative Chaos. Suspicion of utopian speculation, like suspicion of any "overly idealistic" theory, springs from the intuition of the origins of the world in the agonal relation to Chaos. Any consideration of theoretical principles (*archai*) is necessarily colored by these ancient cosmological myths in which *archai* function as determining sources of order. The etymological connection between "princes" and "principles," between "rule" and "rulers," is direct and absolutely pertinent to our understanding of how order is established.

The problem with every principled attempt to establish a model for social order is that it is *arbitrary*. Though we generally agree that "in the beginning there was chaos," we do not agree on what principles we should use to bring order out of our chaotic beginnings. What we intuitively share is the sense of primordial chaos, not the sense of principled order. Thus the problem of social theory in Anglo-European culture is, ultimately, the problem of authority—or its absence. Because in our tradition we are conditioned to believe that order must be authored, the failure to discover the proper author or authors results in a collapse in the sense of author-ity and the resort to arbitrary political power to guarantee minimal social order. Laid bare, our social and political visions—utopian and non-utopian alike—are so many desperate attempts to build dams against the dark, formless waters of primal Chaos. Seen in this context, sanguine utopists have merely applied a somewhat thicker veneer than have their classical colleagues over the perversities

---

[1]Plato did not, of course, invent these principles, nor did the *physiologoi* who preceded him. They existed in embryonic form in Homer who characterized the human personality in terms of *psyche*, *noos*, and *thymos* construed as organs of "life, perception, and of emotion." Bruno Snell, *The Discovery of the Mind*, trans. T.G. Rosemeyer (New York: Harper and Row, 1960), p. 15.

to which, as Children of Chaos, we are willy nilly, held subject.

If utopian speculation is to provide a real alternative to classical social theory, it must be radically an-archic, denying the necessity of principles as external determining sources of order. Taoism offers such an alternative. What distinguishes Taoist forms of anarchism from Anglo-European varieties is revealed in a comparison of the cosmological myths of the two traditions. In our tradition, principles (*archai*) are "sources," or "origins," which serve as the beginnings of thought or action or feeling. These "beginnings"[2] are associated with the time "in the beginning" when Chaos (conceived as confusion, separation, or emptiness) was "overcome." In Taoism, however, there is no fundamental dependence upon creation myths; cosmogonies do not serve as a foundation for social and political order. The Western attitude toward Chaos and the Taoist conception of the coming of "order" out of Chaos differs sharply from that in the West:

> The Emperor of the South Sea was called *Shu* (Brief), the Emperor of the North Sea was called *Hu* (Sudden), and the Emperor of the central region was called *Hun-tun* (Chaos). *Shu* and *Hu* from time to time came together for a meeting in the territory of *Hun-tun*, and *Hun-tun* treated them very generously. *Shu* and *Hu* discussed how they could repay his kindness. "All men," they said, "have seven openings so they can see, hear, eat, and breathe. But *Hun-tun* alone doesn't have any. Let's try boring him some!"
> Every day they bored another hole, and on the seventh day *Hun-tun* died.[3]

Difficult for us to understand is the sympathy for Lord *Hun-tun*. The death of Chaos is the beginning of "order," but an order which is contrived or artificial. Before order, before organization, there was Chaos, but this is "generous," positive harmony, not negative disorder. For the Taoist, *Hun-tun* is best understood as undifferentiated homogeneity," "the sum of all orders."[4] Chaos is the source, origin, or beginning of *all possible* orders. Thus for the Taoist, there is no privileged order. Whereas in the West, the fundamental cosmological principle can be expressed in the proposition that "The Actual World is One," for the Taoist the actual "world" is many. For the Anglo-Europeans, creation is

[2] For a discussion of the etymologies of the terms "beginning" and "principle," and the philosophical significance of their relations, see Chapter Two ("Disciplining Chaos") from my book *The Uncertain Phoenix: Adventures Toward a Post-Cultural Sensibility*, forthcoming from Fordham Univ. Press.

[3] *The Complete Works of Chuang Tzu*, trans. Burton Watson (New York: Columbia Univ. Press, 1968), p. 97.

[4] See Joseph Needham's *Science and Civilization in China*, II (New York: Cambridge Univ. Press, 1956), pp. 107–15.

construal of order from out of a negative Chaos, and "the Beginning" is the beginning of an ordered unity authored by God or the gods. And of order it may be said, "It is good." The Taoist rejoinder is that there is no-beginning in this sense. Chaos is the natural condition of things. The organization of Chaos through human intervention in this natural condition is evil. The principal meaning of "evil" is that which is associated with artificial forms of order or organization.

Taoism locates the essential concern of human existence in the transcendence of order and the return to Chaos as undifferentiated homogeneity. Accordingly, the Taoist conception of psyche, or personality structure, is in dialectical contrast to the Western notion. For the Taoist the functions analogous to thinking, acting, and feeling are expressed in the terms *wu-chih* (no-knowledge), *wu-wei* (non-assertive action) and *wu-yü* (objectless desire). This no-soul doctrine depends upon an essential insight into the character of the world as a Pluriverse. This insight is expessed in two notions—*te* and *tzu-jan*. The former may be translated "intrinsic excellence"; the latter means something like "spontaneity," "naturalness," "self-so."[5] Similarly, the epistemologies of the two traditions are fundamentally different. In our classical tradition, knowing involves insight into the principles of order which characterize the Totality as a Universe, and the application of these principles to particular instances of that orderedness. Thus, to claim that we have knowledge of, say, a natural object—a Tree, for example—is to say that we understand it in terms of its principles—and these may be its essence (*ousia*) expressed as an abstract universal, as a law of organic functioning construed in terms of purpose or aim, as an outcome of material and causal interactions, or simply as a convenient concept. We "know" it by virtue of principles as determining sources of order. In Taoism, on the other hand, knowledge is characterized as "knowing" the intrinsic excellence (*te*) of an object or event. One "knows" a tree through a grasp of its idiosyncratic and unique significance. *Wu-chih*, or no-knowledge, is the consequence of an intuitive grasp of an item as "self-so" (*tzu-jan*), as a spontaneous and self-creating event. No principles as external determining sources of order obtain. Everything has its own "principle," and Nature is just a name for the totality of things or events understood each on its own terms. Such knowledge is no-knowledge in the sense that it does not involve a construal of objects or events in terms of anything other than the events or objects themselves.

Clearly, in Taoism the sharp distinction between nature and culture

[5] A more complete discussion of the significance of these Taoist terms may be found in two of my recent articles: "Process and Anarchy—A Taoist Vision of Creativity," in *Philosophy East and West*, 28, No. 3 (1978), 272—85, and "Praxis, *Karman*, and Creativity," *Philosophy East and West*, 30, No. 1 (1980), 57—64.

to which we are accustomed in the West is no longer relevant. Moreover, Nature, in which all existents find their home, is the sum of all intrinsic excellences and is in no way ordered by external sources of harmony. Thus Taoism may be said to be anarchistic in the purest sense of the term. In the West neither our educative nor naturalistic communitarian notions of social existence are anarchistic in this sense. The former is not because it seeks to establish a culture over and against nature; the latter cannot be since, from a truly anarchistic perspective, there can be no single-ordered nature in accordance with which human beings may seek their common role, function, or purpose.

II. Technology and Anarchy

The very possibility of new forms of social and political theory is in great part the result of technological developments which may permit us to alter our traditional relationship to scientific and moral principles. As a complex of processes, modern technology represents the wedding of the scientific and moral sensibilities. Identified increasingly with both *physis* and *praxis*, technology thus pervades our natural and moral ambience. The rational and moral principles externalized in the objective structures of machines, superhighways, communications networks, and institutions no longer must be part of our psychic equipment, our "inside," but may become, instead, human environs. The mind is freed: in the modern world, it is no longer necessary to expend most of one's energy in establishing and maintaining instrumental relations with the external world.

As objectified structures, technologies *in-form* the world in which we live. Much of the discussion of the developments in information and communication technologies in contemporary society fails to mention this most significant consequence of advanced technology—*viz, informing* our ambience mitigates the requirement that we ourselves be so informed. Instrumental understandings, like instrumental actions, may be turned over to technical, rather than human, instruments. Even the most grandiose of instrumental understandings comprised by religious and scientific cosmologies which promote visions of the world meant to defend us against the fearful chaos of human contingencies may be eschewed. These traditional "sacred canopies" (Peter Berger) are being replaced by *secular* canopies comprised by the sustaining and mediating institutions of our technical environs. The technological in-formation of our world frees us to become informed in ways not heretofore conceived.

We need not allow ourselves to be blinded to the positive possibilities attendant upon this type of social analysis by the nay-saying of those cultural prophets who, in a quaintly archaic vocabulary drawn from an already passing industrial society, proclaim economic and political doctrines painfully irrelevant to complex technological culture. "Alienation," a principal problem of industrial society, becomes, in this technological age, a primary goal of the technical order. We need no longer be hypnotized by the possible alienating consequences of interaction with our technologies if we take it upon ourselves to "alienate" our technical ambience (i.e., to "transfer" it; to "turn it over" to its own self-augmenting devices). For the rational and moral sensibilities institutionalized in and through technology are simply not among the *inalienable* aspects of a fully human existence. Freed from the necessities of mere instrumental understandings and actions we may, perhaps, begin to inform ourselves by what is truly essential for our self-actualizations.

In our technological society, the externalization of moral and scientific sensibilities has institutionalized, formalized, and routinized activities which formerly were left to human initiative. Two important, interrelated consequences of this concern the interplay between technology and politics. First, the existence of a public sphere of action in which there is significant discussion and debate and in which the consequences of decisions can be assessed is one of the conditions of viable political activity. Hannah Arendt, attuned to the classical features of political activity and the importance of tradition in the characterization of the modern age, has claimed that one of the main reasons for the decline in the prestige of politics in the contemporary period is the loss of the public sphere and the increased privatization of social existence.[6] The challenge to traditional politics comes not only from the fact that private satisfactions are preferred to the rewards of public life, but also from increased demands for technical efficiency to meet the more pressing problems of complex societies. The immanent rationale of technologies is the efficient utilization of time and resources. But politics, traditionally, is anything but efficient. Indeed, the personal exercise of power—the *raison d'être* of the politician—is, potentially, among the most inefficient of human activities. For this reason, any alliance between the technician and the politician in contemporary complex societies must be an extremely uncomfortable one.

Secondly, our technological and political future depends largely upon the kinds of social problems we will have to face. If these are primarily ecological dilemmas, we shall require technocratic solutions. Tensions of international relations, on the other hand, require the talents of the power politician. In such a situation military technologies will assume the stage and our political future will in all likelihood be highly

---

[6]See Hannah Arendt, *The Human Condition* (Garden City, New York: Doubleday and Co., Inc., 1959).

unstable. In what follows I shall be taking "utopian license" by assuming the better of these two plausible futures.

Assuming that in the future we resort increasingly to technocratic solutions to social problems, we can expect an automatization of the political arena and a consequent decline in the personal exercise of power. Turning from political solutions to technical resolutions of social problems might just mean that we shall be able to avoid *1984*. Whether we shall avoid *Brave New World* is, of course, quite another question. The very move toward increased dependence on private satisfactions rather than public duties could lead us into a behaviorally engineered dystopia of the type Huxley predicted. This is not as likely as it initially appears, however. The intrinsic aim of the technician is efficiency. His extrinsic goals—whether he will construct gadgets of weapons—have traditionally been a function of political and economic elites which determine the character of the marketplace. If, with increasing automatization, the politician is forced to abdicate his public position, the determination of the extrinsic goals of the technical class will be left largely to economically motivated elites. Without a "public" realm of some kind, however, there is no basis for the maintenance of a common set of interests, values, or desires. Individuals seeking a variety of satisfactions in the private sphere may not be amenable to general or over-arching forms of manipulation or control of either the political or the economic variety.

Any society—even the rather strange kind of technocratic order I have been sketching—is comprised of four principal interacting elements: a material order, a complex of laws or regulations, power and its exercise, and ideals which serve as the defining characteristics of social development. The objective or material order of a society constitutes the *thingliness* of the social world. A characteristic of technological society is that the *use* of things receives greater emphasis than their *possession*. This is due not only to economic reasons associated with planned obsolescence or to the production of novel "necessities" aimed at increasing the variety and number of marketable items, but more importantly to the "de-materialization" of the social world (technology exists primarily in the form of means and processes rather than material consequences).

As regards the element of *law*, technological processes permit the increased regularization of decision-making procedures affecting economic, military, and strictly political activities. Just as the de-emphasis upon objects as *property* alters our relationship to the material order, so the automatization of decision procedures associated with the formulation and implementation of laws and regulations affects political activity by challenging the exercise of discretionary power. Power and its exercise are affected radically since, as a consequence of these developments, neither the power that comes from real wealth (economic power), nor the power that derives from political office, can be sustained in their traditional forms under these conditions. The decreased efficacy of purely economic or political interests lessens the possibility of any kind of centralized control over the public order of society.

The final major component of social systems is comprised of the ideals or aims which serve as defining characteristics of social activity. In our technological culture the source of these ideals is radically different. Neither political, economic, nor intellectual elites serve as important sources of social aims or ideals. The over-arching aim of a technological society is the greatest possible efficiency. The particular aims or goals are functions of the particular social problems technology is called upon to solve. Most of our pressing problems are forced upon us by brute circumstance. Neither the problems, nor the proper means of solving them, are of our choosing. Moreover, technological solutions to problems themselves produce immanent aims and values. Jacques Ellul has termed this the "self-augmenting" nature of advancing technology: each technical process gives rise to conditions whose problems and solutions are themselves suggested by the given process.

Granted that there are radical changes wrought by the technological organization of society, is it at all plausible to assume that technology's challenge to traditional forms of social order increases the relevance of the anarchist vision I have sketched above? To most it must seem more likely that technology will promote totalitarianism. The dangers of employing technologies for totalitarian ends must not be overlooked, of course, but the more likely consequences of a forced liaison between politics and technology will be "anarchy" in the decidedly negative sense of that term.

In our complex societies political elites are faced with the impossible task of adjudicating such a vast number of divergent interests and demands that they cannot do so in any efficacious manner. The alternatives to any important decision are so numerous, the consequences so significant, that one cannot afford the risk that such decisions entail. The "optimization" of the alternative consequences of political decisions requires complex analyses which only specialists and their computers can undertake. Moreover, today the "political" activity of our elected and appointed officials more often concerns issues of the private spheres of existence rather than the public realm. As far as public life is concerned, it either is guided by the "muddily-do" principle[7] or else follows the vector of technical decisions. Our public realm is a "waste

---

[7] I borrow this from Kurt Vonnegut who, in a short sentence, grasps the essence of all determinist forms of *realpolitik* in the axiom which grounds his vision of ethics and politics:
We do, doodily do,
What we must, muddily must.

wide anarchy of chaos," to quote Milton, a melange of interests, ideals, demands, and causal determinations.

If I am correct, however, in believing that the principal social consequences of technology are the shrinking of the public sphere and the widening of the private and personal sphere (which is not only a consequence of increased leisure but also the collapse of community in the political and religious sense), perhaps we may see the other, positive form of anarchy realized. An interesting consequence of the increased concern over the unique character of the private sphere has been the recent popular attempts to "de-politicize" it. Authors of works with titles such as *Sexual Politics, The Politics of the Family, The Politics of Inner Space,* and *The Politics of God,* are not, as might first be supposed, acceding to the political character of the private realm. On the contrary, by raising political encroachments in the personal sphere to the level of consciousness, the hope is to de-politicize it and prepare it for the exercise of self-actualization; this desire is grounded in the admittedly vague, but nonetheless significant, intuition that the potentialities for individual satisfaction exist primarily in this sphere.

Furthermore, technology fosters positive anarchy by increasing the freedom of the individual to determine the nature and character of his own personal existence. Technical apparatus rules the public sphere of existence by regularizing human interaction in the area of transportation, communication, economic resources, health services, etc., but such external "principles" or rules of order need not regulate the most meaningful areas of social activity. The technical matrix, the New Nature which forms the first layer of our human eco-system, can be construed as a protective ambience in which what is most truly human can be encouraged.

Conceived in this fashion, technological society may serve as an appropriate model of an anarchistic utopia. This model of a techno-verse suggests that our social world is not ordered singly through the imposition of external principles but is the sum of all orders individually realized within the private sphere. The social world need no longer be structured primarily in accordance with economic and political elites which define the norms of social interactions. Abstract principles of right and wrong, or notions of duty construed in terms of obedience to law, may no longer be as essential in characterizing the individual's engagement with the social realm. Nor, for that matter, need we have recourse to standardized concepts of beauty and ugliness, which only serve to moralize and trivialize the aesthetic sense and thus to impoverish both private and public enjoyments.

III. Irony and Anarchy

If we were to give a name to the prevailing mood which characterizes our individual and social response to the world in which we live, we should, perhaps, recalling Matthew Arnold, name it "high moral seriousness." Our primary sense of the world is that it is, indeed a *serious affair*. The literalist attitude expressed in the classical scientific search for univocal meanings on which to ground our understandings of the nature of things, and the dogmatic attitude expressed by moralists who wish to articulate and promulgate true ethical principles as a means of structuring the realm of human praxis, are the two fundamental forms in which our mood of high moral seriousness has been directly expressed.

In its most general character, this seriousness is a response to the perception of our existence as radically contingent. A principal source of this perception is the notion of *creatio ex nihilo*. Our world, such is the belief derived primarily from our Hebraic heritage, is the consequence of an arbitrary act of a Divine Agent, a Supreme *Arché*, who is the author of our existence. This *Arché* may be either the Absolute Will celebrated in *Genesis*, or a Divine Mind patterned by the Eternal Ideas of Plato, or some other functional equivalent. It implies a serious commitment to maintaining a single-ordered world founded in accordance with a principle, or system of principles, to which we are rationally and morally bound.

The anarchist utopia considered in this essay challenges the fundamental assumptions which give rise to this mood of high moral seriousness. Seen from the perspective of this utopic model, the technological society is not a fragmented parody of a single normative social system, but rather a "mosaic society" in which the fragments of existence are in fact the positive elements of our social world. Experienced as a positive chaos, such a world is necessarily fragmented and allusory, each item in it reflecting and giving meaning to every other, with no single point of view providing an ultimate focus of significance. In a techno-verse the realm of social praxis is primarily an individualized, privatized sphere in which each individual possesses an intrinsic inalienable excellence.

If we can see our world in these terms, we need not "take it seriously." Since there is no Supreme *Arché* as source of meaning and value, there is no need to "set the world right." Such an *ironic* sensibility will accept the tensions and contradictions that exist at the heart of our experience of the world. For unless we can free ourselves from the bias that forces us to conceive the Cosmos as a single-ordered world which it is our responsibility to recreate in social and political dimensions, we shall surely not escape the temptation to exploit the instrumental power of our complex technologies for totalitarian ends. Taking the world too seriously

will lead us to despair of the relativity of action, thought, and circumstance which gives our present societies such a disordered cast. If we do
despair we shall be seduced into believing that *any* order is better than
negative Chaos, the only alternative to posited order the serious-minded
can envision.

The Implosion of Meaning in the Media and The Implosion of the Social in the Masses

JEAN BAUDRILLARD

W e live in a world of proliferating information and shrinking sense.[1]

Let us consider three hypotheses:

1. Either information produces meaning (the negentropic factor), but wages a losing battle against the constant drain of sense which is taking place on all sides. The wastage even outstrips any effort to reinject meaning through the media, whose faltering powers must be bolstered up by appealing to a productivity of the base. We are dealing with the ideology of free speech, of media reduced to innumerable individual broadcasting units—"anti-media" even.

2. Or, information has nothing to do with meaning. It is qualitatively different, another kind of working model which remains exterior to meaning and its circulation, properly speaking. This is Shannon's hypothesis: that a purely instrumental sphere of information exists which implies no absolute meaning and which cannot itself, therefore, be implicated in any value judgment. It is rather a sort of code, similar perhaps to the genetic code, simply itself, while sense is an entirely different thing, an after-effect, as it is for

---

[1] Except for this instance, where, for the sake of euphony, I have translated it as "sense," I have preferred "meaning" as a rendering for the French *sens.—Trans.*

Monod in *Le Hasard et la nécessité*.[2] Were this so, there would be no significant relationship between the inflation of information and the dwindling of meaning.

3. Or, on the contrary, there is a strict and necessary correlation between the two, to the extent that information destroys or at least neutralizes sense and meaning, the loss of which is directly related to the dissuasive and corrupt action of media-disseminated information.

This last option is the most interesting, but it runs counter to received opinion. Socialization is measured everywhere by the degree of exposure to mediated messages, hence underexposure to the media is believed to make for a de-socialized or virtually a-social individual. Information is everywhere supposed to produce an accelerated circulation of meaning, which appreciates in value as a result, just as capital appreciates as a result of accelerated turnover. Information is presented as being generative of communication, and in spite of the extravagant waste involved, the consensus is that over all a residue of meaning persists, which redistributes itself among the interstices of the social fabric, just as, according to the consensus, material production, in spite of its dysfunctions and irrationality, results, nonetheless, in increased wealth and a more highly developed society. We all pander to this myth, the alpha and omega of our modernity, without which the credibility of our social organization would collapse. The fact is, however, that it is already collapsing, and for this very reason, because whereas we believe that information produces meaning, communication, and the "social," it is exactly the opposite which obtains.

The social is not a clear and unequivocal process. The question arises whether modern societies are the result of a process of progressive socialization, or *de*-socialization? Everything depends on the meaning of the term socialization, whose various meanings are unstable, even reversible.

Thus, for all the institutions which have marked the "progress" of the social (urbanization, centralization, production, work, medicine, education, social security, insurance, including capital itself, doubtless the most powerful medium of socialization), it could be claimed that they at once produce and destroy the social.

If the social is made up of abstract demands which arise, one after the other, on the ruins of the ritual and symbolic edifice of earlier societies, then these institutions produce more and more of them. But at the same time they sanction that wasting abstraction, whose specific target is

---

[2]Jacques Monod, *Le Hasard et la nécessité* (Paris: Editions du Seuil, 1970).

perhaps, the very marrow of the social. From this point of view one could say that the *social* regresses in direct proportion to the development of its institutions.

This process accelerates and reaches its peak in the case of the mass media and information. *All* the media and *all* information cut both ways: while appearing to augment the social, in reality they neutralize social relations and the social itself, at a profound level.

Information devours its own content; it devours communication and the social, for two reasons:

1. Instead of facilitating communication, it exhausts itself in the *staging* of communication. Instead of producing meaning, it wears itself out staging it. This is the gigantic simulation-process with which we are familiar: non-directive interviews, phone-ins, all-round participation, verbal blackmail—"you're involved, the event is you," etc. The domain of information is increasingly invaded by this kind of phantom content, a homeopathic graft, a fantasy of communication. It is a circular arrangement by which the audience's desire is staged, an anti-theater of communication, which, as we know, is never anything more than the recycling of the traditional institution in negative form, the integrated circuit of the negative. What energy is expended in order to keep this sham at arm's length, to avoid the brutal de-simulation which would bring us face to face with the reality of a radical loss of meaning!

It is futile to wonder whether it is the loss of communication which puts this sham at a premium, or whether the pretense is there from the start, fulfilling a dissuasive purpose: that of short-circuiting in advance any possibility of communication, a precession of the model which eradicates the real. It is futile to wonder what the first term is, there is none—it is a circular process, that of simulation, of the hyperreal. Hyperreality of communication and meaning: by dint of being more real than the real itself, reality is destroyed.

Thus both the social and communication function in a closed circuit like a decoy which assumes the power of a myth. Faith in the existence and value of information is confirmed by that tautological proof that the system provides of itself, by reduplicating, through the medium of signs, a reality that is in fact undiscoverable, chimeric.

But what if this faith is just as ambiguous as that which was inspired by myth in archaic societies? One both believes it and doesn't believe it at the same time, without questioning it seriously, an attitude that may be summed up in the phrase: "Yes, I know, but all the same." There is a kind of reverse shamming among the masses, and in each of us, at the individual level, which corresponds to that travesty of meaning and communication in which we are imprisoned by the system. The tautology of the system is met by ambivalence, dissuasion by disaffection or by a belief that is at least enigmatic. The myth exists, but one must beware of thinking that people believe in it: that is the trap of critical thought, which can only operate on the assumption that the masses are naive and stupid.

2. Behind this exaggerated staging of communication, the mass media, the continuous build-up of information, pursue their relentless destructuring of the *social*.

A bombardment by signs, which the masses are supposed to echo back, an interrogation by converging light/sound/ultra-sound waves, linguistic or light stimuli exactly like distant stars, or the nuclei that are bombarded by particles in a cyclotron: that is information. Not a mode of communication or of meaning but a state of perennial emulsion, of input-output and controlled chain reactions, such as exists in atomic simulation chambers. The "energy" of the mass must be liberated in order to be transformed into the "social."

It is a contradictory process, however, because information, in all its forms, instead of intensifying or even creating the "social relationship" is, on the contrary, an entropic process, a modality of the extinction of the social.

The idea is that the masses are structured, their captive social energy liberated, by the injection of information and messages (it is not so much the circumscribing action of the institutions, as the quantity of information received and the rate of exposure to the media which is the measure of socialization today). But the opposite is true. Instead of transforming the mass into energy, information simply produces more and more mass. Instead of informing, as it claims, that is to say, conferring form and structure, information increasingly neutralizes the "social field," creating ever larger inert masses, impervious to classic social institutions, to the very content of information itself. The nuclear explosion of symbolic structures by the social, and its rational violence, is succeeded today, by the fission of the social itself, by the irrational violence of the media and information—the final result being precisely an atomized, nuclearized, molecularized mass.

All that is left are fluid, mute masses, the variable equations of surveys, objects of perpetual tests in which, as in an acid solution, they are dissolved. Testing, probing, contacting, soliciting, informing—these are tactics of microbiological warfare undermining the social by infinitesimal dissuasion, so that there is no longer even time enough for crystallization to take place. Hitherto, violence used to crystallize the social, violently bringing forth an antagonistic social energy: a repressive demiurgy. Today it is a gentle *semi*urgy which controls us.

Thus information dissolves meaning and the social into a sort of nebula and is committed, not to increased innovation, but on the contrary, to total entropy.

I have so far dealt with information only in the social sphere of communication. It would be intresting, however, to carry the hypothesis into the domain of cybernetic information theory. There too, the obtaining thesis holds that cybernetic information is synonymous with negentropy, resistance to entropy, increased meaning, and improved organization. But it behooves us to pose the inverse hypothesis: Information = Entropy. For example: the information or knowledge which it is possible to have about a system or an event is already a form of neutralization and entropy of that system. This is true of the sciences in general and of the human and social sciences in particular. The information which reflects or diffuses an event is already a degraded form of that event. The role of the media in May, 1968 is a case in point. The coverage given to the student revolt gave rise to the general strike, but the latter turned out to be precisely a neutralizing black box, an antidote to the initial virulence of the movement. The publicity itself was a death trap. One must be wary of the attempt to universalize strategies through the dissemination of information, suspicious of all-out campaigns for a solidarity which is both electronic and fashionable. Every strategy which is geared to the universalization of differences is an entropic strategy of the system.

Thus the media do not facilitate socialization, but rather the implosion of the social in the masses. This is simply the macroscopic extension of the implosion of meaning which occurs at the microscopic level of the sign. Marshall McLuhan's "The Medium is the Message" (the consequences of which are far from having been exhausted) is the tool for further analysis of this situation. By this phrase, McLuhan means that all subject matter (the message) is absorbed by the single dominant form of the medium. The medium alone is the event, regardless of whether the message it conveys is conformist or subversive, a state of affairs which poses a serious problem for all forms of counter information, pirate radio stations, anti-media, etc. But there are more serious consequences to come, which McLuhan himself has not defined. For even if meanings are neutralized, the possibility might still exist of working on the form of the medium so that its purely formal impact might transform the real. Devoid of meaning, the medium *as such* might still retain a revolutionary, subversive use-value. *But* (and this is the point to which McLuhanism, if pursued to the limit, leads) it is not simply a question of the implosion of the message in the medium, but of the medium and the real in a kind of hyperreal nebula where the very definition and distinctive action of the medium is irrecoverably lost. In a word, "The Medium is the Message" does not merely mean the end of the message, but the end of the medium. There are no more media, in the literal sense of the term (I am referring particularly to the electronic mass media), that is to say, in the sense of mediating between one state of reality and another, and that is true for both form and content. That, strictly speaking, is what implosion means: the defusing of polarities, the short-circuiting of the poles of every differential system of meaning, the obliteration of distinctions and oppositions between terms, including the distinction between the medium and the real. Hence any mediation or dialectical intervention between the two, or by one on the other, becomes impossible. We are faced with the circularity of all media-effects. Meaning too, in the sense of a unilateral vector leading from one pole to another, becomes impossible. The consequences of this critical and unprecedented situation must be confronted; it is the only one left to us and it is futile to dream of a revolution through either form or content, since both the medium and the real now form a single inscrutable nebula.

This statement (of the implosion of content, of the consumption of meaning, of the evanescence of the medium itself, of the re-absorption of any dialectic of communication in the absolute circularity of the model, of the implosion of the social in the masses) might appear to be a prophecy of doom. This is so only in relation to the idealism which colors our thinking about information. We are all nourished by an exaggerated idealism of meaning and communication (meaning idealizes communication) and from that perspective, it is the catastrophe of meaning which lies in wait.

It is necessary to understand, however, that the word catastrophe does not carry the catastrophic connotation of death and dissolution except in a linear perspective of accumulation, of productive finality which the system imposes on us. Etymologically, the term itself merely indicates the curve, the downward turn towards the bottom of the cycle, which leads to what might be called "the horizon of the event," an insurmountable horizon of meaning, beyond which nothing further can occur *which has meaning for us*. But it suffices to escape from that ultimatum of meaning for catastrophe itself to drop the appearance of the final, nihilistic deadline which it appears to us to be now. Beyond meaning lies fascination: the result of the neutralization and implosion of meaning. Beyond the horizon of the social, there are masses, which result from the neutralization and implosion of the social. Is not the opposition of fascination and meaning what is at stake in information?

Whatever its content, be it political, pedagogical, cultural, the objective of information is always to circulate meaning, *to subjugate the masses to meaning*. This imperative to produce meaning translates itself into an impulse to moralize: to inform better, to socialize better, to raise the cultural level of the masses, etc. What nonsense! The masses remain scandalously resistant to this imperative of rational communication. They are offered meaning when what they want is entertainment. The best efforts have failed to get them to focus on the seriousness of the subject matter, or even of the code. They are given messages when they want only the sign. They delight in the interplay of signs and stereotypes and in any content, as long as it results in a dramatic sequence. What they reject is the "dialectic" of meaning. And it serves no purpose to claim that they are mystified. This is a hypocritical hypothesis designed merely to protect the intellectual comfort of the producers of meaning, according to which the masses are simply prevented from enjoying the natural light of reason. Such a hypothesis serves to exorcise its opposite: that it is in *perfect freedom* that the masses prefer entertainment to meaning and the ultimatum it poses. They do not trust that transparence and that *political* intent an inch. They intuit the terrorizing simplification behind the ideal hegemony of meaning and they react in their own way, by reducing all articulate discourse to a single irrational, groundless dimension, in which signs lose their meaning and subside into exhausted fascination.

It is a question of their own demand, of a specific and positive counter-strategy, which consists in absorbing and obliterating culture, knowledge, power, and the social: an operation which has been going on since time immemorial, but never before on such a scale. What is at issue is a deeply embedded antagonism which forces a reversal of the given scenario. It is no longer *meaning* which constitutes the ideal line of force in our society (in which case whatever escapes meaning is regarded as a kind of detritus to be reabsorbed at one time or another) but rather meaning itself which is merely an accident, ambiguous and transitory, an effect of the ideal convergence of perspectives (of history, of power, etc.) at a given moment. But this event has, in fact, never affected more than a tiny fraction, a very superficial stratum of our society either collectively or individually. We are but intermittently purveyors of meaning. For the most part, we form a mass in the fullest sense of the word, living panic-stricken lives, subject to the laws of chance, straddling meaning, never coinciding with it.

For example, on the night of Klaus Croissant's extradition, there was a TV broadcast of a football game in which France was playing for the World Cup. A few hundred people demonstrated in front of the Santé prison, there was some furious nocturnal activity on the part of a few lawyers, while twenty million people spent the evening in front of their TV screens. The joy at France's victory was equalled only by the dismay and outrage which such scandalous indifference inspired among the more enlightened. As *Le Monde* put it: '9 O'Clock. The German lawyer has already been removed from La Santé prison. In a few minutes Rocheteau will score the first goal." Thus, melodramatic indignation, but not a single question as to the cause of this mysterious indifference. The reason advanced for it is always the same: the manipulation of the masses by the powers that be, the mystifying effect of football. In any case, the suggestion is that such indifference *ought not* to exist, and has, therefore, nothing to say to us. In other words, "the silent majority" is dispossessed even of its indifference, which is not even recognized and correctly attributed; thus its very apathy is not spontaneous, but prompted by the powers that be.

What contempt lies behind this interpretation! Victims of mystification as they are, how could one expect the masses to know how to behave? A certain revolutionary spontaneity, through which they glimpse the "rationality of our own desire" is occasionally conceded to them, but Heaven protect us from their silence and inertia! It is precisely this indifference, however, which needs to be analyzed in its *positive* brutality, rather than consigning it to a white magic of alienation and manipulation which constantly deflects the masses from their natural calling. But how does this mumbo-jumbo succeed? Should we not ask ourselves why, after several revolutions and one or two centuries of political apprenticeship, in spite of newspapers, unions, political parties, intellectuals, and all the energy invested in educating and mobilizing the people, it is still the case, and will still be the case in ten or twenty years, that while a thousand are willing to take a stand, twenty million are content to remain passive, and not simply passive, but actively preferring, joyfully and unquestioningly, in good faith, a football match to an event of grave political and human consequence? It is indeed curious that this phenomenon has served, not to disturb the existing analysis, but rather to confirm its belief in the omnipotence of manipulation and of the masses as supine and comatose, neither of which has any foundation in reality and both of which are decoys. Power manipulates nothing, and the masses are neither led astray nor mystified. The powers that be are all too happy to place on football an easy responsibility, if not the diabolical responsibility for the stultification of the masses. This reinforces them in the illusion of *being* the powers that be, and diverts their attention from a much more dangerous fact, to wit that this indifference of the masses is their sole, their true practice, that there is no other imaginable ideal, that it is not to be deplored, but rather to be analyzed as the brute fact of collective denial, the refusal to accept the ideals, however luminous, which are put before them.

The masses have no stake in these ideals. We might as well acknowledge the fact and recognize that every aspiration of the social, every hope of revolution and social change up till now has functioned only by virtue of this subterfuge, this blind spot, this incredible memory lapse. We might as well do what Freud attempted in the psychic domain, that is to take this remainder, this blank sediment, this detritus of meaning, this unanalyzed and perhaps unanalyzable matter, as our point of departure.

The masses are represented in the imaginary as floating somewhere between passivity and barbarous spontaneity, the constant source of potential energy, a reserve of the social and social energy, today a silent referent, tomorrow the protagonist in history, once they have spoken out and ceased to be the "silent majority." The truth is, however, that the masses do not have a history to write, either past or future, they have no potential energy to release or desires to fulfill: their power is completely present, in the here and now. *It is the power of their silence.* They possess a power of absorption and neutralization already greater than any external force, a power of specific inertia whose efficacy is different from all our power systems, from all the schemata of production and expansion on which our imaginary functions, even when it seeks their destruction. This is the inadmissible and unintelligible figure of implosion (is it even a process?): the thrust of all our meaning-systems, against which they muster their defences as they attempt to cover up with a tissue of signification the crater left by the central collapse of meaning.

Every majority has not always been silent, but today, it is so by definition. Perhaps it has been *reduced* to silence, but this is by no means certain. Because this silence, if it means that the majority does not speak, indicates above all that it is no longer possible to speak in its name. No one represents the silent majority, or the masses. We can no longer refer to them, as we did formerly to "class" or to the "people." Silent and withdrawn, the masses are no longer a *subject* (certainly not of history); they can no longer, therefore, be spoken, articulated, represented, cannot pass through the political mirror-stage nor the cycle of imaginary identifications. The power resulting from this is obvious, because, no longer a subject, *the masses are no longer capable of alienation,* either in their own language (they do not have one) or in any other which would presume to speak for them. This puts an end to revolutionary expectations, which have always gambled on the possibility of the masses or a given class denying itself as such. The masses, however, are a locus not of negativity and explosion, but of absorption and implosion.

That is the paradox of this silence. While appearing to be the ultimate manifestation of alienation, it is the ultimate weapon. The masses are impervious to the workings of liberation, revolution, or historicity, but that is their method of self-defense, their riposte. They are a simulation model, an alibi at the service of a phantom political class which no longer knows the nature of the political power it exerts over the masses, and at the same time, the death of that very political process which is supposed to govern them. The masses swallow up the political, insofar as it implies will and representation.

Critical thought judges, discriminates, and produces differences. It is by this selection process that it acts as the guardian of meaning. The masses, on the other hand, do not choose, do not produce differences but indifference; they preserve the fascination of the medium, which they prefer to the critical demands of the message. For fascination does not stem from meaning, it is rather exactly proportionate to the alienation of meaning. It establishes itself by favoring the medium over the message, the idol over the idea, the simulacrum over the truth. It is on this level that the media function. Fascination is their law and the violence that is specific to them, a brutal violence perpetrated on meaning, which cancels out communication via meaning in favor of another mode. The question is, which one?

The hypothesis that communication might take place outside the medium of meaning, that the very intensity of the communication might be directly proportionate to the reabsorption and dissolution of meaning, seems to us untenable. Yet it is not meaning, or the increase of meaning that produces intense pleasure. It is rather its neutralization that fascinates us. (Cf. my discussion of the joke, der Witz.[3]) And this is not the result of any death wish—which would imply that "life" is still on the side of meaning—but simply a challenge to reference, message, code, to all the categories of the linguistic enterprise, all of which are disavowed in favor of an implosion of the sign in fascination: no more signifiers or signifieds but a reabsorption of the poles of signification. None of the watch-dogs of meaning can understand that. Meaning is morally outraged by fascination.

What is essential today is to evaluate the double challenge presented to meaning—that of the masses and their silence (by no means passive)—that of the media and their fascination. All marginal, alternative attempts to reawaken meaning must remain subordinate to that. Clearly there is a paradox at the basis of the close link of the masses with the media. Is it the media which dissolve meaning and produce the formless or "informed" masses, or is it the masses which defeat the media by perverting or silently absorbing all the messages which they produce? I have analyzed and condemned the media as the institution of an irreversible model of one-way communication, in "Requiem pour les media."[4] But what of today? The failure of the masses to respond to the media may be interpreted not as the strategy of power, but as a counter-strategy of the masses themselves, vis-à-vis power. What then?

Are the mass media on the side of power in the manipulation of the masses or are they on the side of the masses in the liquidation of meaning, in the violence done to meaning, in the mechanism of fascination? Is it the

[3]Jean Baudrillard, L'Echange symbolique et la mort (Paris: Gallimard, 1976).
[4]Jean Baudrillard, "Requiem pour les media," in Pour une critique de l'économie politique du signe (Paris: Gallimard, 1972), pp. 220–28.

media that cast the spell of fascination on the masses, or the masses who reduce the media to spectacle? The Mogadishu-Stammheim incidents indicate that the media condemn terrorism and the exploitation of fear for political ends, at the same time that they diffuse, completely ambiguously, the brutal fascination of the terrorist act.[5] This is the eternal moral dilemma which Umberto Eco posits: how not to speak of terrorism, how to put the media to good use, when such good use is impossible. The media convey sense and nonsense in equal measure, they manipulate every which way, and the process is uncontrollable. They transmit both the simulation which is internal to the system and that which would destroy it according to an entirely circular and moebian logic—and that is as it should be. There is no alternative, no logical resolution, only a logical exacerbation and a catastrophic resolution.

With this qualification: we are in a double-bind situation vis-à-vis the system, exactly as children are when they encounter the demands of the adult world. Children are simultaneously required on the one hand, to establish themselves as autonomous, responsible, free, conscious subjects, and on the other, as inert, submissive, obedient, and conforming objects. The child resists on every front and meets contradictory demands with a double strategy. To the demand that he be an object he opposes every variety of disobedience, revolt, emancipation: in a word, the claim of the subject. He counters the demand that he be a subject just as obstinately and effectively with a reverse strategy, that is to say, infantilism, hyper-conformism, total dependence, passivity, idiocy: the resistance of the object. Neither of these two strategies has more objective value than the other. Today, however, the resistance of the subject is unilaterally valorized just as in the political sphere only liberation, emancipation, free expression, all constitutive of the political subject, are held to be valuable and subversive. This is to ignore the equal, doubtless even greater, impact of the object-strategies, of renunciation of the position of the subject and of meaning—precisely the strategy of the masses, which we bury contemptuously under the labels of alienation and passivity. Liberating practices deal with only one side of the system, the ultimatum to make ourselves objects. They ignore the other demand, that we be subjects: liberate ourselves, express ourselves, vote, produce, decide, talk, participate, play the game, although this demand represents

[5]The reference is to the hijacking of a Lufthansa plane with eighty-six passengers on board by West German terrorists in October 1977. The West German government refused the terrorists' demand that eleven members of the Bader-Meinhoff gang be freed in exchange for the hostages, who were rescued by commandos when the latter re-captured the plane at the airport of Mogadishu, the capital of Somalia. Ulrike Meinhoff had been found hanged in prison at Stammheim in May 1976.

as serious an ultimatum, as serious a blackmail, as the other, perhaps even more so at the present time. The strategic resistance to a system based on oppression and repression is the liberating claim of the subject. But this is to respond to an anterior phase of the system, which, even if it confronts us still, is no longer the center of operations. The current strategy of the system is to inflate utterance[6] to produce the maximum of meaning. Thus the appropriate strategic resistance is to refuse meaning and utterance, to simulate in a hyper-conformist manner the very mechanisms of the system, itself a form of refusal and non-reception. This is the resistance strategy of the masses. It amounts to turning the system's logic back on itself by duplicating it, reflecting meaning, as in a mirror, without absorbing it. This is the dominant strategy at the present (if one can call it strategy) because it is this particular phase of the system that has triumphed.

To choose the wrong strategy is a grave error. Any movement which stakes everything on the liberation, the emancipation, the resurrection of a historical, collective, or speaking subject, on a raising of consciousness if not of the unconscious, individually and collectively, is blind to the fact that it is conforming to the system, whose goal at the present time is precisely the overproduction and regeneration of meaning and utterance.

Translated by Mary Lydon
University of Wisconsin-Milwaukee

---

[6]The word used by Baudrillard is *parole*: notoriously difficult, with its Saussurian resonance, to render in English. I have opted for "utterance" because it carries the connotation both of speech and of individual use of language.—*Trans.*

III
Art and Technology:
The Avant Garde

The Hidden Dialectic:
The Avant Garde—Technology—
Mass Culture

ANDREAS HUYSSEN

> Historical materialism wishes to retain
> that image of the past which unexpect-
> edly appears to man singled out by
> history at a moment of danger. The
> danger affects both the content of the
> tradition and its receivers. The same
> threat hangs over both: that of becoming
> a tool of the ruling classes. In every era
> the attempt must be made anew to wrest
> tradition away from conformism that is
> about to overpower it.
>
> Walter Benjamin,
> *Theses on the Philosophy of History*

I

When Walter Benjamin, one of the foremost theoreticians of avant garde art and literature, wrote these sentences in 1940 he certainly did not have the avant garde in mind at all. It had not yet become part of that tradition which Benjamin was bent on salvaging. Nor could Benjamin have foreseen to what extent conformism would eventually overpower the tradition of avant gardism, both in advanced capitalist societies and, more recently, in East European societies as well.

151

Like a parasitic growth, conformism has all but obliterated the original iconoclastic and subversive thrust of the historical avant garde[1] of the first three or four decades of this century. This conformism is manifest in the vast depoliticization of post-World War II avant garde art and its institutionalization as administered culture,[2] as well as in academic interpretations which, by canonizing the historical avant garde, modernism and postmodernism, have methodologically severed the vital dialectic between the avant garde and mass culture in industrial civilization. In most academic criticism the avant garde has been ossified into an elite enterprise beyond politics and beyond everyday life, though their transformation was once a central project of the historical avant garde.

In light of the tendency to project the post-1945 depoliticization of culture back onto the earlier avant garde movements, it is crucial to recover a sense of the cultural politics of the historical avant garde. Only then can we raise meaningful questions about the relationship between the historical avant garde and the neo-avant garde, modernism and postmodernism, as well as about the aporias of the avant garde and the consciousness industry (Hans Magnus Enzensberger), the tradition of the new (Harold Rosenberg) and the death of the avant garde (Leslie Fiedler). For if discussions of the avant garde do not break with the oppressive mechanisms of hierarchical discourse (high vs. popular, the new new vs. the old new, art vs. politics, truth vs. ideology), and if the question of today's literary and artistic avant garde is not placed in a larger sociohistorical framework, the prophets of the new will remain locked in futile battle with the sirens of cultural decline—a battle which by now only results in a sense of déjà vu.

II

Historically the concept of the avant garde, which until the 1930s was not limited to art but always referred to political radicalism as well,[3] assumed prominence in the decades following the French Revolution. Henri de Saint Simon's *Opinions littéraires, philosophiques et industrielles* (1825) ascribed a vanguard role to the artist in the construction of the ideal state and the new golden age of the future,[4] and since then the concept of an

[1]The term "historical avant garde" has been introduced by Peter Bürger in his *Theorie der Avantgarde* (Frankfurt am Main: Suhrkamp, 1974). It includes mainly dadaism, surrealism and the post-revolutionary Russian avant garde. See Bürger, p. 44.

[2]Cf. Theodor W. Adorno, "Culture and Administration," *Telos*, 37 (Fall 1978), 93–111.

[3]See Donald Drew Egbert, *Social Radicalism and the Arts: Western Europe* (New York: Alfred A. Knopf, 1970).

[4]On the authorship of relevant sections of the *Opinions*, see Matei Calinescu, *Faces of Modernity: Avantgarde, Decadence, Kitsch* (Bloomington: Indiana Univ. Press, 1977), p. 101f.

avant garde has remained inextricably bound to the idea of progress in industrial and technological civilization. In Saint Simon's messianic scheme, art, science, and industry were to generate and guarantee the progress of the emerging technical-industrial bourgeois world, the world of the city and the masses, capital and culture. The avant garde, then, only makes sense if it remains dialectically related to that for which it serves as the vanguard—speaking narrowly, to the older modes of artistic expression, speaking broadly, to the life of the masses which Saint Simon's avant garde scientists, engineers, and artists were to lead into the golden age of bourgeois prosperity.

Throughout the nineteenth century the idea of the avant garde remained linked to political radicalism. Through the mediation of the utopian socialist Charles Fourier, it found its way into socialist anarchism and eventually into substantial segments of the bohemian subcultures of the turn of the century.[5] It is certainly no coincidence that the impact of anarchism on artists and writers reached its peak precisely when the historical avant garde was in a crucial stage of its formation. The attraction of artists and intellectuals to anarchism at that time can be attributed to two major factors: artists and anarchists alike rejected bourgeois society and its stagnating cultural conservatism, and both anarchists and left-leaning bohemians fought the economic and technological determinism and scientism of Second International Marxism, which they saw as the theoretical and practical mirror image of the bourgeois world.[6] Thus, when the bourgeoisie had fully established its domination of the state and industry, science and culture, the avant gardist was not at all in the forefront of the kind of struggle Saint Simon had envisioned. On the contrary, he found himself on the margins of the very industrial civilization which he was opposing and which, according to Saint Simon, he was to prophesy and bring about. In terms of understanding the later condemnations of avant garde art and literature both by the right (*entartete Kunst*) and by the left (bourgeois decadence), it is important to recognize that as early as the 1890s the avant garde's insistence on cultural revolt clashed with the bourgeoisie's need for cultural legitimation, as well as with the preference of the Second International's cultural politics for the classical bourgeois heritage.[7]

[5] See Egbert, *Social Radicalism and the Arts;* on bohemian subcultures, see Helmut Kreuzer, *Die Boheme* (Stuttgart: Metzler, 1971).

[6] See David Bathrick and Paul Breines, "Marx und/oder Nietzsche: Anmerkungen zur Krise des Marxismus," in *Karl Marx und Friedrich Nietzsche,* ed. R. Grimm and J. Hermand (Königstein/Ts.: Athenäum, 1978).

[7] See Andreas Huyssen, "Nochmals zu Naturalismus-Debatte und Linksopposition," in *Naturalismus—Ästhetizimus,* ed. Christa Bürger, Peter Bürger, and Jochen Schulte-Sasse (Frankfurt am Main: Suhrkamp, 1979).

Neither Marx nor Engels ever attributed major importance to culture (let alone avant garde art and literature) in the working-class struggles, although it can be argued that the link between cultural and political-economic revolution is indeed implicit in their early works, especially in Marx's Parisian Manuscripts and the *Communist Manifesto*. Nor did Marx or Engels ever posit the Party as the avant garde of the working class. Rather, it was Lenin who institutionalized the Party as the vanguard of the revolution in *What Is to Be Done* (1902) and soon after, in his article "Party Organization and Party Literature" (1905), severed the vital dialectic between the political and cultural avant garde, subordinating the latter to the Party. Declaring the artistic avant garde to be a mere instrument of the political vanguard, "a cog and screw of one single great Social Democratic mechanism set in motion by the entire politically conscious avantgarde of the entire working class,"[8] Lenin thus helped pave the way for the later suppression and liquidation of the Russian artistic avant garde which began in the early 1920s and culminated with the official adoption of the doctrine of socialist realism in 1934.[9]

In the West, the historical avant garde died a slower death, and the reasons for its demise vary from country to country. The German avant garde of the 1920s was abruptly terminated when Hitler came to power in 1933, and the development of the West European avant garde was interrupted by the war and the German occupation of Europe. Later, during the cold war, especially after the notion of the end of ideology took hold, the political thrust of the historical avant garde was lost and the center of artistic innovation shifted from Europe to the United States. To some extent, of course, the lack of political perspective in art movements such as abstract expressionism and Pop art was a function of the altogether different relationship between avant garde art and cultural tradition in the United States, where the iconoclastic rebellion against a bourgeois cultural heritage would have made neither artistic nor political sense. In the United States, literary and artistic heritage never played as central a role in legitimizing bourgeois domination as it did in Europe. But these explanations for the death of the historical avant garde in the West at a certain time, although critical, are not exhaustive. The loss of potency of the historical avant garde may be related more fundamentally to a broad cultural change in the West in the twentieth century: it may be argued that the rise of the Western culture industry, which paralleled the decline of the historical avant garde, has made the avant garde enterprise itself obsolete.

[8]V.I. Lenin, 'The Re-organization of the Party' and 'Party Organization and Party Literature' (London: IMG Publications, 1972), p. 17.
[9]See Hans-Jürgen Schmidt and Godehard Schramm, eds., *Sozialistische Realismuskonzeptionen: Dokumente zum 1. Allunionskongress der Sowjetschriftsteller* (Frankfurt am Main: Suhrkamp, 1974).

To summarize: since Saint Simon, the historical avant garde of Europe had been characterized by a precarious balance of art and politics, but since the 1930s the cultural and political avant gardes have gone their separate ways. In the two major systems of domination in the contemporary world, the avant garde has lost its cultural and political explosiveness and itself become a tool of legitimation. In the United States, a depoliticized cultural avant garde has produced largely affirmative culture, most visibly in pop art where the commodity fetish reigns supreme. In the Soviet Union and in Eastern Europe, the historical avant garde was first strangled by the iron hand of Stalin's cultural henchman Zhdanov and then revived as part of the cultural heritage, thus providing legitimacy to regimes which face growing cultural and political dissent.

Both politically and aesthetically, today it is important to retain that image of the now lost unity of the political and artistic avant garde, which may help us forge a new unity of politics and culture adequate to our own times. Since it has become more difficult to share the historical avant garde's belief that art can be crucial to a transformation of society, the point is not simply to revive the avant garde. Any such attempt would be doomed, especially in a country such as the United States where the European avant garde failed to take roots precisely because no belief existed in the power of art to change the world. Nor, however, is it enough to cast a melancholy glance backwards and indulge in nostalgia for the time when the affinity of art to revolution could be taken for granted. The point is rather to take up the historical avant garde's insistence on the cultural transformation of everyday life and from there to develop strategies for today's cultural and political context.

III

The notion that culture is a potentially explosive force and a threat to advanced capitalism (and to bureaucratized socialism, for that matter) has a long history within Western Marxism from the early Lukács up through Habermas's *Legitimation Crisis* and Negt/Kluge's *Öffentlichkeit und Erfahrung*.[10] It even underlies, if only by its conspicuous absence, Adorno's seemingly dualistic theory of a monolithically manipulative culture industry and an avant garde locked into negativity. Peter Bürger, a recent theoretician of the avant garde, draws extensively on this critical Marxist tradition, especially on Benjamin and Adorno. He argues convincingly that the major goal of art movements such as Dada, surrealism, and the post-1917 Russian avant garde was the reintegration of art into life praxis,

---

[10]Jürgen Habermas, *Legitimation Crisis*, trans. Thomas McCarthy (Boston: Beacon Press, 1975); Oskar Negt and Alexander Kluge, *Öffentlichkeit und Erfahrung* (Frankfurt am Main: Suhrkamp, 1972).

the closing of the gap separating art from reality. Bürger interprets the widening gap between art and life, which had become all but unbridgeable in late nineteenth-century aestheticism, as a logical development of art within bourgeois society. In its attempt to close the gap, the avant garde had to destroy what Bürger calls "institution art," a term for the institutional framework in which art was produced, distributed, and received in bourgeois society, a framework which rested on Kant's and Schiller's aesthetic of the necessary autonomy of all artistic creation. During the nineteenth century the categorical separation of art from reality and the insistence on the autonomy of art, which had once freed art from the fetters of church and state, had worked to push art and artists to the margins of society. In the art for art's sake movement, the break with society—the society of imperialism—had led into a dead end, a fact painfully clear to the best representatives of aestheticism. Thus the historical avant garde attempted to transform l'art pour l'art's isolation from reality—which reflected as much opposition to bourgeois society as Zola's j'accuse—into an active rebellion that would make art productive for social change. In the historical avant garde, Bürger argues, bourgeois art reached the stage of self-criticism; it no longer only criticized previous art qua art, but also critiqued the very "institution art" as it had developed in bourgeois society since the eighteenth century.[11]

Of course, the use of Marxian categories of criticism and self-criticism implies that the negation and sublation (Aufhebung) of the bourgeois "institution art" is bound to the transformation of bourgeois society itself. Since such a transformation did not take place, the avant garde's attempt to integrate art and life almost had to fail. This failure, later often labelled the death of the avant garde, is Bürger's starting point and his reason for calling the avant garde "historical." And yet, the failure of the avant garde to reorganize a new life praxis through art and politics resulted in precisely those historical phenomena which make any revival of the avant garde's project today highly problematic, if not impossible: namely, the false sublations of the art/life dichotomy, in fascism with its aesthetization of politics,[12] in Western mass culture with its fictionalization of reality, and in socialist realism with its claims of reality status for its fictions.

If we agree with the thesis that the avant garde's revolt was directed against the totality of bourgeois culture and its psycho-social mechanisms of domination and control, and if we then make it our task to salvage the historical avant garde from the conformism which has obscured its

[11]See Bürger, Theorie der Avantgarde, esp. chapter 1.

[12]See Rainer Stollman, "Fascist Politics as a Total Work of Art: Tendencies of the Aesthetization of Political Life in National Socialism," New German Critique, 14 (Spring 1978), 41–60.

political thrust, then it becomes crucial to answer a number of questions which go beyond Bürger's concern with the "institution art" and the structure of the avant garde art work. How precisely did the dadaists, surrealists, futurists, constructivists, and productivists attempt to overcome the art/life dichotomy? How did they conceptualize and put into practice the radical transformation of the conditions of producing, distributing, and consuming art? What exactly was their place within the political spectrum of those decades and what concrete political possibilities were open to them in specific countries? In what way did the conjunction of political and cultural revolt inform their art and to what extent did that art become part of the revolt itself? Answers to these questions will vary depending on whether one focusses on Bolshevik Russia, France after Versailles, or Germany, doubly beaten by World War I and a failed revolution. Moreover, even within these countries and the various artistic movements, differentiations have to be made. It is fairly obvious that a montage by Schwitters differs aesthetically and politically from a photomontage by John Heartfield, that Dada Zurich and Dada Paris developed an artistic and political sensibility which differed substantially from that of Dada Berlin, that Mayakovsky and revolutionary futurism cannot be equated with the productivism of Arvatov or Gastev. And yet, as Bürger has convincingly suggested, all these phenomena can legitimately be subsumed under the notion of the historical avant garde.

IV

I will not attempt here to answer all these questions, but will focus instead on uncovering the hidden dialectic of avant garde and mass culture, thereby casting new light on the objective historical conditions of avant garde art, as well as on the socio-political subtext of its inevitable decline and the simultaneous rise of mass culture.

Mass culture as we know it in the West is unthinkable without twentieth-century technology—media techniques as well as technologies of transportation (public and private), the household, and leisure. Mass culture depends on technologies of mass production and mass reproduction and thus on the homogenization of difference. While it is generally recognized that these technologies have substantially transformed everyday life in the twentieth century, it is much less widely acknowledged that technology and the experience of an increasingly technologized life world have also radically transformed art. Indeed, technology played a crucial, if not *the* crucial, role in the avant garde's attempt to overcome the art/life dichotomy and make art productive in the transformation of everyday life. Bürger has argued correctly that from dada on the avant garde movements distinguish themselves from preceding movements such as impressionism, naturalism, and cubism not only in their attack on the "institution art" as such, but also in their radical break with the referential mimetic aesthetic and its notion of the autonomous and organic work of art. I would go further: no other single factor has influenced the emergence of the new avant garde art as much as technology, which not only fueled the artists' imagination (dynamism, machine cult, beauty of technics, constructivist and productivist attitudes), but penetrated to the core of the work itself. The invasion of the very fabric of the art object by technology and what one may loosely call the technological imagination can best be grasped in artistic practices such as collage, assemblage, montage and photomontage; it finds its ultimate fulfillment in photography and film, art forms which can not only be reproduced, but are in fact designed for mechanical reproducibility. It was Walter Benjamin who, in his famous essay "The Work of Art in the Age of Mechanical Reproduction," first made the point that it is precisely this mechanical reproducibility which has radically changed the nature of art in the twentieth century, transforming the conditions of producing, distributing, and receiving/consuming art. In the context of social and cultural theory Benjamin conceptualized what Marcel Duchamp had already shown in 1919 in L.H.O.O.Q. By iconoclastically altering a reproduction of the Mona Lisa and, to use another example, by exhibiting a mass-produced urinal as a fountain sculpture, Marcel Duchamp succeeded in destroying what Benjamin called the traditional art work's aura, that aura of authenticity and uniqueness which constituted the work's distance from life and which required contemplation and immersion on the part of the spectator. In another essay, Benjamin himself acknowledged that the intention to destroy this aura was already inherent in the artistic practices of dada.[13] The destruction of the aura, of seemingly natural and organic beauty, already characterized the works of artists who still created individual rather than mass-reproducible art objects. The decay of the aura, then, was not as immediately dependent on techniques of mechanical reproduction as Benjamin had argued in the Reproduction essay. It is indeed important to avoid such reductive analogies between industrial and artistic techniques and not to collapse, say, montage technique in art or film with industrial montage.[14]

It may actually have been a new experience of technology that sparked the avant garde rather than just the immanent development of the artistic forces of production. The two poles of this new experience of

[13]See Walter Benjamin, "The Author as Producer," in Understanding Brecht, trans. Anna Bostock (London: New Left Books, 1973), p. 94. See also Andreas Huyssen, "The Cultural Politics of Pop," New German Critique, 4 (Winter 1975), 83–84, 89ff.

[14]See Burkhardt Lindner and Hans Burkhard Schlichting, "Die Dekonstruktion der Bilder: Differenzierungen im Montagebegriff," alternative, 122/123 (Oct./Dec. 1978), 218–21.

technology can be described as the aesthetization of technics since the late nineteenth century (world expos, garden cities, the *cité industrielle* of Tony Garnier, the *Città Nuova* of Antonio Sant'Elia, the *Werkbund*, etc.) on the one hand and the horror of technics inspired by the awesome war machinery of World War I on the other. And this horror of technics can itself be regarded as a logical and historical outgrowth of the critique of technology and the positivist ideology of progress articulated earlier by the late nineteenth-century cultural radicals who in turn were strongly influenced by Nietzsche's critique of bourgeois society. Only the post-1910 avant garde, however, succeeded in giving artistic expression to this bipolar experience of technology in the bourgeois world by integrating technology and the technological imagination in the production of art.

The experience of technology at the root of the dadaist revolt was the highly technologized battlefield of World War I—that war which the Italian futurists glorified as total liberation and which the dadaists condemned as a manifestation of the ultimate insanity of the European bourgeoisie. While technology revealed its destructive power in the big *Materialschlachten* of the war, the dadaists projected technology's de-structivism into art and turned it aggressively against the sanctified sphere of bourgeois high culture whose representatives, on the whole, had enthusiastically welcomed the war in 1914. Dada's radical and disruptive moment becomes even clearer if we remember that bourgeois ideology had lived off the separation of the cultural from industrial and economic reality, which of course was the primary sphere of technology. Instrumental reason, technological expansion, and profit maximization were held to be diametrically opposed to the *schöner Schein* (appearance of beauty) and *interesseloses Wohlgefallen* (disinterested pleasure) dominant in the sphere of high culture.

In its attempt to reintegrate art and life, the avant garde of course did not want to unite the bourgeois concept of reality with the equally bourgeois notion of high, autonomous culture. To use Marcuse's language, they did not want to weld the reality principle to affirmative culture, since these two principles constituted each other precisely in their separation. On the contrary, by incorporating technology into art, the avant garde liberated technology from its instrumental aspects and thus undermined both bourgeois notions of technology as progress and art as "natural," "autonomous," and "organic." On a more traditional representational level, which was never entirely abandoned, the avant garde's radical critique of the principles of bourgeois enlightenment and its glorification of progress and technology were manifested in scores of paintings, drawings, sculptures, and other art objects in which humans are presented as machines and automatons, puppets and mannequins, often faceless, with hollow heads, blind or staring into space. The fact that these presentations did not aim at some abstract "human condition," but rather critiqued the invasion of capitalism's technological instrumentality into the fabric of everyday life, even into the human body, is perhaps most evident in the works of Dada Berlin, the most politicized wing of the Dada movement. While only Dada Berlin integrated its artistic activities with the working-class struggles in the Weimar Republic, it would be reductive to deny Dada Zurich or Dada Paris any political importance and to decree that their project was "only aesthetic," "only cultural." Such an interpretation falls victim to the same reified dichotomy of culture and politics which the historical avant garde had tried to explode.

V

In Dada, technology mainly functioned to ridicule and dismantle bourgeois high culture and its ideology, and thus was ascribed an iconoclastic value in accord with dada's anarchistic thrust. Technology took an entirely different meaning in the post-1917 Russian avant garde—in futurism, constructivism, productivism, and the proletcult. The Russian avant garde had already completed its break with tradition when it turned openly political after the revolution. Artists organized themselves and took an active part in the political struggles, many of them by joining Lunacharsky's NARKOMPROS, the Commissariat for Education. Many artists automatically assumed a correspondence and potential parallel between the artistic and political revolution, and their foremost aim became to weld the disruptive power of avant garde art to the revolution. The avant garde's goal to forge a new unity of art and life by creating a new art and a new life seemed about to be realized in revolutionary Russia.

This conjunction of political and cultural revolution with the new view of technology became most evident in the LEF group, the productivist movement, and the proletcult. As a matter of fact, these left artists, writers, and critics adhered to a cult of technology which to any contemporary radical in the West must have seemed next to inexplicable, particularly since it expressed itself in such familiar capitalist concepts as standardization, americanization, and even taylorization. In the mid-1920s, when a similar enthusiasm for technification, Americanism, and functionalism had taken hold among liberals of the Weimar Republic, George Grosz and Wieland Herzfelde tried to explain this Russian cult of technology as emerging from the specific conditions of a backward agrarian country on the brink of industrialization and rejected it for the art of an already highly industrialized West: "In Russia this constructivist romanticism has a much deeper meaning and is in a more substantial way socially conditioned than in Western Europe. There constructivism is partially a natural reflection of the powerful technological offensive of the beginning industrialization."[15] And yet, originally the technology cult was more than just a reflection of industrialization, or, as Grosz and Herzfelde also suggest, a propagandistic device. The hope that people such as Tatlin, Rodchenko, Lissitsky, Meyerhold, Tretyakov, Brik, Gastev, Arvatov, Eisenstein, Vertov, and others invested in technology was closely tied to the revolutionary hopes of 1917. With Marx they insisted on the qualitative difference between bourgeois and proletarian revolutions. Marx had subsumed artistic creation under the general concept of human labor, and he had argued that human self-fulfillment would only be possible once the productive forces were freed from oppressive production and class relations. Given the Russian situation of 1917, it follows almost logically that the productivists, left futurists, and constructivists would place their artistic activities within the horizon of a socialized industrial production: art and labor, freed for the first time in history from oppressive production relations, were to enter into a new relationship. Perhaps the best example of this tendency is the work of the Central Institute of Labor (CIT), which, under the leadership of Aleksej Gastev, attempted to introduce the scientific organization of labor (NOT) into art and aesthetics.[16] The goal of these artists was not the techno-logical development of the Russian economy at any cost—as it was for the Party from the NEP period on, and as it is manifest in scores of later socialist realist works with their fetishization of industry and technology. Their goal was the liberation of everyday life from all its material, ideological, and cultural restrictions. The artificial barriers between work and leisure, production and culture were to be eliminated. These artists did not want a merely decorative art which would lend its illusory glow to an increasingly instrumentalized everyday life. They aimed at an art which would intervene in everyday life by being both useful and beautiful, an art of mass demonstrations and mass festivities, an activating art of objects and attitudes, of living and dressing, of speaking and writing. Briefly, they did not want what Marcuse has called affirmative art, but rather a revolutionary culture, an art of life. They insisted on the psycho-physical unity of human life and understood that the political revolution could only be successful if it were accompanied by a revolution of everyday life.

VI

In this insistence on the necessary "organization of emotion and

[15]George Grosz and Wieland Herzfelde, "Die Kunst ist in Gefahr. Ein Orientierungs-versuch" (1925), cited in Diether Schmidt, ed., Manifeste—Manifeste. Schriften deutscher Künstler des 20. Jahrhunderts, I (Dresden: n.d.), 345–46.

[16]See Karla Hielscher, "Futurismus und Kulturmontage," alternative, 122/123 (Oct./Dec. 1978), 226–35.

thought" (Bogdanov), we can actually trace a similarity between late nineteenth-century cultural radicals and the Russian post-1917 avant garde, except that now the role ascribed to technology has been totally reversed. It is precisely this similarity, however, which points to interesting differences between the Russian and the German avant garde of the 1920s, represented by Grosz, Heartfield, and Brecht among others.

Despite his closeness to Tretyakov's notions of art as production and the artist as operator, Brecht never would have subscribed to Tretyakov's demand that art be used as a means of the emotional organization of the psyche.[17] Rather than describing the artist as an engineer of the psyche, as a psycho-constructor,[18] Brecht might have called the artist an engineer of reason. His dramatic technique of *Verfremdungseffekt* relies substantially on the emancipatory power of reason and on rational ideology critique, principles of the bourgeois enlightenment which Brecht hoped to turn effectively against bourgeois cultural hegemony. Today we cannot fail to see that Brecht, by trying to use the enlightenment dialectically, was unable to shed the vestiges of instrumental reason and thus remained caught in that other dialectic of enlightenment which Adorno and Horkheimer have exposed.[19] Brecht, and to some extent also the later Benjamin, tended toward fetishizing technique, science, and production in art, hoping that modern technologies could be used to build a socialist mass culture. Their trust that capitalism's power to modernize would eventually lead to its breakdown was rooted in a theory of economic crisis and revolution which, by the 1930s, had already become obsolete. But even there, the differences between Brecht and Benjamin are more interesting than the similarities. Brecht does not make his notion of artistic technique as exclusively dependent on the development of productive forces as Benjamin did in his Reproduction essay. Benjamin, on the other hand, never trusted the emancipatory power of reason and the *Verfremdungseffekt* as exclusively as Brecht did. Brecht also never shared Benjamin's messianism or his notion of history as an object of construction.[20] But it was especially Benjamin's emphatic notion of experience (*Erfahrung*) and profane illumination which separated him from Brecht's enlightened trust in ideology critique and pointed to a definite affinity between Benjamin and the Russian avant garde. Just as Tretyakov, in his futurist poetic strategy, relied on shock to alter the

[17]Sergej Tretjakov, *Die Arbeit des Schriftstellers: Aufsätze, Reportagen, Porträts* (Reinbek: Rowohlt, 1972), p. 87.

[18]Tretjakov, p. 88.

[19]Max Horkheimer and Theodor W. Adorno, *Dialectic of Enlightenment* (New York: Herder & Herder, 1972).

[20]See Walter Benjamin, "Theses on the Philosophy of History," in *Illuminations*, ed. Hannah Arendt (New York: Schocken Books, 1969).

psyche of the recipient of art, Benjamin, too, saw shock as a key to changing the mode of reception of art and to disrupting the dismal and catastrophic continuity of everyday life. In this respect, both Benjamin and Tretyakov differ from Brecht: the shock achieved by Brecht's *Verfremdungseffekt* does not carry its function in itself but remains instrumentally bound to a rational explanation of social relations which are to be revealed as mystified second nature. Tretyakov and Benjamin, however, saw shock as essential to disrupting the frozen patterns of sensory perception, not only those of rational discourse. They held that this disruption is a prerequisite for any revolutionary reorganization of everyday life. As a matter of fact, one of Benjamin's most interesting and yet undeveloped ideas concerns the possibility of a historical change in sensory perception, which he links to a change in reproduction techniques in art, a change in everyday life in the big cities, and the changing nature of commodity fetishism in twentieth-century capitalism. It is highly significant that just as the Russian avant garde aimed at creating a socialist mass culture, Benjamin developed his major concepts concerning sense perception (decay of aura, shock, distraction, experience, etc.) in essays on mass culture and media as well as in studies on Baudelaire and French surrealism. It is in Benjamin's work of the 1930s that the hidden dialectic between avant garde art and the utopian hope for an emancipatory mass culture can be grasped alive for the last time. After World War II, at the latest, discussions about the avant garde congealed into that reified two-track system of high vs. low, elite vs. popular, which itself is the historical expression of the avant garde's failure and of continued bourgeois domination.

VII

Today, the obsolescence of avant garde shock techniques, whether dadaist, constructivist, or surrealist, is evident enough. One need only think of the exploitation of shock in Hollywood productions such as *Jaws* or *Close Encounters of the Third Kind* in order to understand that shock can be exploited to reaffirm perception rather than change it. The same holds true for a Brechtian type of ideology critique. In an age saturated with information, including critical information, the *Verfremdungseffekt* has lost its demystifying power. Too much information, critical or not, becomes noise. Not only is the historical avant garde a thing of the past, but it is also useless to try to revive it under any guise. Its artistic inventions and techniques have been absorbed and co-opted by Western mass mediated culture in all its manifestations from Hollywood film, television, advertising, industrial design, and architecture to the aesthetization of technology and commodity aesthetics. The legitimate place of a cultural avant garde which once carried with it the utopian hope for an emancipatory mass culture under socialism has been preempted by the rise of mass mediated culture and its supporting industries and institutions.

Ironically, technology helped initiate the avant garde artwork and its radical break with tradition, but then deprived the avant garde of its necessary living space in everyday life. It was the culture industry, not the avant garde, which succeeded in transforming everyday life in the twentieth century. And yet—the utopian hopes of the historical avant garde are preserved, even though in distorted form, in this system of secondary exploitation euphemistically called mass culture. It seems preferable for cultural theory today to address the contradictions of technologized mass culture rather than pondering over the products and performances of the various neo-avant gardes, which, more often than not, derive their originality from social and aesthetic amnesia. Today the best hopes of the historical avant garde may not be embodied in art works at all, but in decentered movements which work toward the trans-formation of everyday life. The point then would be to retain the avant garde's attempt to address those human experiences which either have not yet been subsumed under capital, or which are stimulated but not fulfilled by it. Aesthetic experience in particular must have its place in this transformation of everyday life, since it is uniquely apt to organize fantasy, emotions, and sensuality against that repressive desublimation which is so characteristic of capitalist culture since the 1960s.

Art and Technology: Alienation or Survival?

MIKEL DUFRENNE

The word *techne* in Greek refers to any productive practice; it does not include distinctions made today under the headings of technique and art. Today the word "art" refers above all to "the way of doing something according to a certain method,"[1] and adjectives are used to differentiate the industrial arts from the liberal arts and to designate the Fine Arts which now claim a monopoly of the Beautiful. As long as beauty was not dogmatically conceived and defined, that which man produced, at least when not pressed by need, was beautiful. Beauty lay in the potential for meaning which realized the object, an object simultaneously useful, integrated and integral, or rather, both the product and production of nature prescribed by a unifying vision of the world, gathering about it and in it the milieu from which it took root. Heidegger's well-known analyses of the thing (*la chose*), be it a vase or a temple, remain pertinent. This natural beauty, the product of the co-naturality of man and nature exclusively, owed nothing to purely artistic practice. But art broke free. Having initiated the split between art and technique, art became progressively autonomous and institutionalized as the artist demanded and obtained a status distinct from that of the artisan; the commercialization of art today guarantees the specificity of works of art and the privilege accorded to genius. But technique has not lagged

[1] Emile Littré, *Dictionnaire de la langue française*, 1956 ed.

165

behind. In an industrial society, as technique develops it becomes institutionalized in turn: parallel to Schools of Fine Arts are Institutes of Technology and Schools of Engineering. Technique tends to be an end in itself. It affirms itself and reflects itself as technology; it exerts a growing power as technocracy. We have entered the technological age without art in return proposing an artology or imposing an artocracy.

To grasp more firmly the relationship which was established after this split between art and technique, we must take a look at what is at stake in the technological age. In terms of the machine, it is important to distinguish what Revault d'Allones calls technical truth from social truth the operation of the machine from its use. With regard to its operation, Gilbert Simondon has clearly pointed out that "the enslaving violence" of technocracy is not involved. The machine needs man, and when the machine takes its place as a technical component in a technical ensemble, a web of individual parts which are interconnected, it calls on man as an "associate," a "witness." Simondon observes: "the relationship between man and machine develops at the level of processes of transduction"; man can organize the technical ensemble in terms of the virtual because he has a "sense of time that the machine does not have because it is not alive."[2] Simondon is perhaps so careful to do justice to the technical object that he considers its operation to be an end in itself, an end to which man is not a means if, precisely, he participates in it, if, according to the technological point of view, "he is not only preoccupied with the use of a technical entity, but also with the correlation of those entities with each other." But Simondon is not unaware that technique, rather than existing for itself, is used as the instrument of an "unbridled will to conquer."[3] He thus denounces the Faustian dream of a conquering aggression which violates nature.

This is a perversion of reason, argue Horkheimer and Adorno, who stress that man's domination of nature is accomplished by man's domination of man—to the point where both the objective of technique (its end is no longer utility but profit and power) and the operation of technique (the worker is kept unaware of what he is doing, he is not a partner but a servile subject) are altered. In order to understand this denaturation of technology by technocracy, the socio-political system must be taken into account; one must, for example, become conversant with the Frankfurt School.

But my intention in this paper is not that. It is rather to focus on the conflicting relationships which have developed between technique and art since the radical difference between the line of the engineer's pencil

[2]Gilbert Simondon, *Du mode d'existence des objets techniques* (Paris: Aubier-Montaigne, 1969), p. 144.
[3]Simondon, p. 145, p. 137.

and the stroke of the artist's brush has become apparent. Technique first. Technique is enough of an imperialist both to deny the tie which links it with art and to claim to dominate art. Either technique reduces art to a subordinate ornamental role, declaring it to be exterior and insignificant, as, for example, chrome is added to an automobile or one per cent of art is injected into public buildings; or technique repudiates art by putting itself in art's place: according to functionalism, the object produced by technique must be self-contained; it is beautiful insofar as it meticulously assumes its technicity; design has nothing to learn from art.

And art? Is it equally intransigent? Art cannot so easily forget the primitive *techne*. For art is above all doing, a setting up as Souriau would say, or poiesis, as Valéry would put it. Doing in the sense of perfecting, with pleasure, with taste. A loving battle with a resisting material, friendship with the tool that extends the body, a flirtation with the obstacle, a game of chance in which one never establishes enough control to eliminate all surprise. Ask the engravers and the potters. They will tell you that what is most authentic in art is the tinkering, and even musicians and architects will not deny it. This tinkering is the hand of technique, which art, of necessity, conceals.

But today? How should art deal with the technique conceptualized by technology, a technique much more powerful, more sure of itself, more authoritative? The answers to this question are very different, depending on the arts and artists involved. Certain arts deal with a material over which technology has little hold, like dance whose material is the body, poetry whose material is language, traditional painting whose material is canvas and pigments. Other arts, on the contrary, are themselves generated by modern technique, such as cinema and video, or better yet the cinematic art that Nicolas Schöffer talks about.[4] And the traditional arts can also be tempted, such as painting which sometimes turns to the computer, or even music which the machine can serve in multiple ways, as Daniel Charles shows, on the one hand, by expanding the acoustical universe through the manipulation of noise, the production of electronic sounds, and the arrangement of the diffusion of sound in space, and on the other hand, by introducing new structures which, with the help of the computer, allow the operation of chance. To these entreaties artists are sensitive in very different ways. A panorama of contemporary art would reveal a wide variety of attitudes, from derision to decision, as Revault d'Allonnes has said, from an obstinate refusal of a radical modernity to a resolute acceptance of it. At one pole Duchamp or Tinguely, all the creators of celibate or desiring machines, at the other pole, Schöffer

[4] See Nicolas Schöffer's essay on "Art et techologie," in *Medianalyses*, ed. Marie-Pierre de Montgoméry (Nice, France: Centre du 20eme Siècle, forthcoming).

or Xenakis. In between the two we find difference, a sweet form of refusal.

These two polar attitudes suggest a freedom which is equal. And equally, both positions can be misunderstood, or at least subject to various interpretations. And this is so because the refusal of modern technique can stem from a loyalty to the traditional practice of art, to the pleasure of creating in accordance with the immemorial gesture of the hand that moves over the keyboard or holds the brush, the engraving tool, the gouge, or the pen, while the eye or the ear experiments and judges; and indeed, it is perhaps in these diligent and patient motions that man feels himself most intensely alive. Or, the refusal can stem from a turning inward which transforms the ivory tower into a private museum where phantasy-objects are collected, apart from the world, apart from time. Yet the refusal of technique, or of a technical work, in its most acute forms—aggression or derision—is first of all a political refusal; it is not a refusal of technology as such, but a refusal of technological civilization which is responsible for the aberrations produced by technique. The invention of a machine which caricatures machines is not a call for their destruction, it is a denunciation of their use in the system of domination; even when the artist stays out of politics, his or her art can be politicized and the message can be received.

What are the implications of the acceptance and use of modern technique? First of all, the desire to befit the times: not yielding to the nostalgia of the past, not choosing flight or exile, but encouraging the development of an art which liberates itself by acknowledging the gains of modernity and using them to invent rather than repeat. For a new world, new art. Enough has been said about the death of art in the world's literature—art would indeed die if it did not renew itself, if it did not take up the challenge which technology presents. Fine, this premise has its merits. I contend, however, that it implies a monolithic view of culture. Totality, even if characterized by conflicts, must be homogeneous, and institutions survive only by being adaptable. But is it not possible for culture to harbour small islands of anarchy? In any case, attached to this acceptance of technology is the double risk for art of enslavement and alienation, because, in fact, in industrial society, technology, however innocent and even benevolent it may be in itself, is in the hands of those who have it at their disposal for their own ends—power and profit. By turning to technology, does art not aid and abet these powers? If so, it would not be for the first time. As soon as a system of domination prevails, art conveys the dominant ideology, brings the powers that be to the forefront and glorifies them. Today would not a cybernetic tower which sings about the city be an instrument of the police? This enfeoffment to the system can be all the more tempting to art in that, by exploiting the resources of technology, art also is in danger of becoming perverted, of losing its playful character and its dedication to pleasure in order to assume a stance and an authority proportionate to the power conferred on it by these very resources.

These risks, however, can be and often are thwarted—the risk of perversion, for example—if artistic practice remains faithful to what has always been its goal, which is the production of an object or an event to be savoured to the point of absorption, inviting play and pleasure. (Let us recall in passing that *jouissance* is not what is displayed by the pornocrats, that pleasure concerns the psyche as much as the body, that it can be full with thoughts and laden with anguish, in short, according to a film of Max Ophuls, *Le Plaisir, ce n'est pas gai*, at least not always.) And for the work to appeal to us in this way, it must be created in the same *Stimmung*, put together with love (*"Le petit saint Paul veut encore deux jours de caresses,"* wrote Poussin to Chanteloup). The question is, can one tinker with a computer? Xenakis answers yes. With a piece of film or an editing table? All of experimental cinema answers yes. Thus the other risk I mentioned is already offset. Because, in short, the question is one of separating technology and technocracy, and what that requires is simply bending technology to a use other than that of those in power, to use it playfully. The appropriation of technique, the abduction of it by putting it to work for music, that is, for pleasure and not for profit, this is how Revault d'Allonnes understands Xenakis's efforts. I quote from d'Allonnes at some length:

> Seen in this light, Xenakis's music is a victory of man over the "cyberna-thrope," the robot. Far from being bound by the computer, Xenakis keeps the computer subservient, subservient to this plan, as human as it can be, of producing a music that is both new and accessible at the same time. One is tempted to say that Xenakis's "production" uses the most advanced technical processes of production tongue in cheek and almost as a challenge to the very processes themselves. He manages to achieve, by his use of these processes, a reversal of the relationship between man and technique traditionally established by different social classes.[5]

Likewise for experimental cinema. It has been said, too quickly, that the movie camera is the bearer of ideology; more precisely, it is the use of the camera by commercial cinema that makes this so. But let a different, a liberated cinema play freely with the same camera, and also with the roll of film, and cinema is immediately restored to innocence, dedicated, perhaps, to our liberation. Thus for nature, as well as for man, technique is no longer enslaving. The artificial no longer stands in opposition to the natural with the purpose of attacking it. Simondon has already shown

[5]Olivier Revault d'Allonnes, *La Création artistique et les promesses de la liberté* (Paris: Klincksieck, 1973), p. 259.

that the technical object "can have its aesthetic epiphany insofar as it extends the world and enters into it"; far from de-naturing nature, it revives in us what Simondon calls the feeling of substance precisely "when it [the object] meets with a suitable substance for which it can truly be the form, that is, when it completes and expresses the world."[6] The lighthouse is to the reef, or the line of pylons to the valley, what the temple is to the city. In addition, the artificial can reveal the natural. Technique lets us hear the song of the dolphin and the noise of the city—why not the music of the spheres? Here technique rejoins art, and oddly enough painting, which does not reproduce the real, but imitates its appearance, this process of appearing by which Nature becomes world.

All things considered, the enterprise of art very likely promotes the self-revelation of technology by subverting its use, by resolutely distinguishing its technical truth from its social truth.[7] "The essence of technique has little to do with technique," says Heidegger.[8] True, and it is perhaps in art that one must look for this essence. But that does not mean that technique should deny itself in as much as self-discovery in the world of instrumentality is limited. It is enough that, torn from the system of domination, technique perpetuates the friendship that art reveals between man and the world.

One last word, one at the heart of the matter. Suppose, and this is a hope encouraged by certain signs today, that the boundaries within which art has confined itself by becoming institutionalized, should crack; that artistic practice should become truly popular, self-administered by the people. Possibly then such a practice would preclude the spread of that advanced technology which is so burdensome. But what does it matter, if art is part of everyone's life? What does it matter if traditional forms, which have not fallen from grace, are perpetuated, if they are taken over by the people who will perhaps transform them? But that is not all. Technical practice itself will be ascribed to art; for this to come about, work must in turn be self-determined; so that, no longer forced, experienced as duty or punishment, but free, it might be experienced as play. For if play implies work, work can include play. Would not this play be a form of art?

Translated by Carol Tennessen
University of Wisconsin-Milwaukee

[6]Simondon, p. 185.

[7]"What joy does humanity owe, up to now, to computers? In the midst of a thousand vexations for which these so-called anonymous administrators, masters, rather than slaves, are responsible, only two blessings, to my knowledge, exist. The first is that they were wrong in Vietnam, and the second is that Xenakis took hold of their technical truth in order to go beyond their social truth" (Revault d'Allonnes, *Xenakis: Les Polytopes* [Paris: Balland, 1975], p. 49.

[8]Martin Heidegger, *Essais et conférences*, trans. André Préau (Paris: Gallimard, 1959), p. 47.

Art and Technics: John Cage, Electronics, and World Improvement

KATHLEEN WOODWARD

> LXXXII.   In music it was hopeless to
> think in terms of the old structure
>   (tonality), to do things following old
>   methods (counterpoint, harmony), to use
> the old materials (orchestral
>   instruments).   We started from
>   scratch: sound, silence, time,
>   activity.   In society, no amount of
> doctoring up economics/politics will help.
>   Begin again, assuming abundance,
>   unemployment, a field situation,
>   multiplicity, unpredictability,
> immediacy, the possibility of
>   participation.
>
>   —John Cage, *Diary: How to Improve the*
>   *World (You Will Only Make Matters*
>   *Worse)*

The most renowned of John Cage's experiments in avant garde music is 4'33" (1952), a timed frame of silence which questions Western man's tradition of imposing his will on art and the environment. In Cage's view, silence is the absence of intention. 4'33" is an ecological piece which instructs us to begin at the root, to listen, quietly, a-new, to silence the restless movement of the mind. As Cage has written, "in

171

music, it was hopeless to think in terms of the old structure. . . . We started from scratch: sound, silence, time, activity."[1]

As in music, so in cultural politics. Aesthetic exercise has been the necessary preparation for action; in recent years Cage has turned to research and writing on how to improve the world:

LXII.   We
open our eyes and ears seeing life
each day excellent as it is.   This
realization no longer needs art though
without art it would have been difficult
(yoga, zazen, etc.) to come by.
Having this realization, we gather
energies, ours and the ones of
nature, in order to make this intolerable
world endurable.   Robots.[2]

For Cage, the method for improving the world is analogous to the method for generating art. "In society," he asserts, "no amount of doctoring up economics/politics will help. Begin again. . . ."[3] And just as technology has been a central concern in his music, so it is in his vision of a future world. As both an artist and cultural critic, he approaches technology in the spirit of fruitful collaboration. Technics do not stand opposed to art. Utopia spins from the circuits of the computer.

This is an uncommon attitude. As David Madden has pointed out, most serious American literature has regarded the machine as a threat to culture.[4] As the work of Lewis Mumford reveals, art and technics have traditionally been seen as antithetical, even antagonistic to one another.[5] But Cage has a romance with technology.[6] To comprehend his enthusiasm, which is tempered by Zen, I will begin by counterpointing Cage's

[1]*Diary: How to Improve the World (You Will Only Make Matters Worse), A Year From Monday* (Middletown, Connecticut: Wesleyan Univ. Press, 1969), p. 157. Unfortunately, it is impossible to reproduce the typefaces and line breaks. For a general introduction to Cage, see Calvin Tomkins' essay in *The Bride and the Bachelors: Five Masters of the Avant-Garde* (New York: Viking Press, 1968). See also *John Cage*, ed. Richard Kostelanetz (New York: Praeger, 1970), and two recent books by the French aesthetician Daniel Charles, *Pour les oiseaux: Entretiens avec Daniel Charles* (Paris: Pierre Belfond, 1977) and *Gloses sur John Cage* (Paris: Union Générale d'Editions, 1978).

[2]*Diary, A Year From Monday*, p. 146.

[3]*Diary, A Year From Monday*, p. 157.

[4]David Madden, "Introduction," *American Dreams, American Nightmares* (Carbondale, Illinois: Southern Illinois Univ. Press, 1970).

[5]Lewis Mumford, *Art and Technics* (New York: Columbia Univ. Press, 1952).

[6]This is Jonathan Benthall's term. See *The Body Electric: Patterns of Western Industrial Culture* (London: Thames and Hudson, 1976).

notions of art and technics with those of Mumford. This will involve distinguishing between industrial and postindustrial technologies and the dominant metaphors or images which characterize a literary response to them. It will also lead us to a brief consideration of the art and technology movement of the sixties. I will then turn to Cage's open-ended *Diary: How To Improve the World (You Will Only Make Matters Worse)* which projects a utopian future based on electronics. Finally, I will evaluate Cage's attitude toward technology in terms of both the substance of the *Diary* and the method of making the text.

I. Art and Technics

In his *Art and Technics* which was published nearly thirty years ago, Mumford declares that technology has gone out of control, repressing the human and impoverishing our inner life. Ideally art should resist the dehumanization wrought by the machine:

> Most of the great artists of the last two centuries—and this has been equally true, I think, in music and poetry and painting, even in some degree architecture—have been in revolt against the machine and have proclaimed the autonomy of the human spirit: its autonomy, its spontaneity, its inexhaustible creativeness.[7]

For Mumford, art is concerned with the expression of the inner self and the forging of social bonds. Art is the articulation of love; technics, the incarnation of power. Art is subjective, technics, objective. Art is organic, technics, mechanical. Exploring the personality is the province of art, mastering nature is the goal of technics. Technics, Mumford writes, is "that part of human activity wherein, by an energetic organization of the process of work, man controls and directs the forces of nature for his own purposes"; technics is "concerned primarily with the enlargements of human power" in man's external environment.[8] Impersonality, regularity, repetition, mechanical accuracy, efficiency, reliability, uniformity: these are the values associated with technics. As Mumford sees it, the great problem of our time is to balance the claims of art and technics, to restore man's wholeness by reasserting the needs of the inner life and regenerating man's symbol-making powers. Our obsession with the machine must be broken, the mechanical stranglehold on our life loosened.[9]

[7]Mumford, p. 7.
[8]Mumford, p. 15, p. 24.
[9]But we must not oversimplify Mumford's position. Although he does oppose the two systems of value embodied in art and technology, he does not believe that they must necessarily be in conflict. Furthermore, although Mumford separates art and technics for theoretical reasons at the beginning of his book, he admits later that "in actual history, this

Mumford's view of the conflict between art and technology in the modern machine age is a familiar one. The words "modern" and "machine" are not chosen casually. In *Art and Technics*, Mumford refers to the development of modern technology primarily in terms of "mechaniza-tion." His basic touchstone is the technology of the industrial revolution, what has come to be called the machine age. Lately, however, we have heard much talk of a *post*mechanical, *post*industrial, *post*modern age which is founded primarily on a technology which produces services, not manufactured goods. Industrial technology is said to be a technology based on power; postindustrial technology, one based on information. Industrial technology is embodied in machines which replace animal or muscle power, postindustrial technology in machines which extend the nervous system or brain.[10] Distinguishing between mechanical and electrical-electronic technologies, the philosopher of technology Hans Jonas writes:

> As mechanics was the first form in which natural science had emerged, so the first stage of technology ushering in the industrial revolution was what we may call the *mechanical* stage. Its products were machines made of rigid parts and powered by the mechanics of volume expansion under heat—thus operating with the familiar solids and forces and on the familiar dynamical principles of classical mechanics. Their predominant use was in the manu-facturing of goods and their transportation.[11]

The goods created by mechanical technology and the needs served by it are, Hans Jonas believes, essentially "natural." In contrast, electrical-electronic technologies mark a genuinely new phase in the scientific-technological revolution because they have increased the degree of

---

separation does not hold. Art and technics go together, sometimes influencing each other, sometimes merely having a simultaneous effect upon the worker of the user" (p. 58). He notes that for a long span of human history, "the tool and the object, the symbol and the subject, were not in fact separated. All work was performed directly by human hand" (p. 60). And finally, Mumford's advice to be silent, to learn to pause and listen, is similar to Cage's:

The renewal of life is the great theme of our age, not the further dominance, in ever more frozen and compulsive forms, of the machine. And the first step for each of us is to seize the initiative and to recover our own capacity for living; to detach ourselves sufficiently from the daily routine to make ourselves self-respecting, self-governing, persons. In short, we must take things into our own hands. Before art on any great scale can redress the distortions of our lop-sided technics, we must put ourselves in the mood and frame of mind in which art becomes possible, as either creation or re-creation: above all, we must learn to pause, to be silent, to close our eyes and wait.

[10]See Bernard Gendron, *Technology and the Human Condition* (New York: St. Martin's Press, 1977), pp. 22–39, for a neat mapping of the basic distinctions among the agricultural, industrial, and postindustrial revolutions.

[11]Hans Jonas, *Philosophical Essays: From Ancient Creed to Technological Man* (Englewood Cliffs, N.J.: Prentice Hall, 1974), p. 75.

artificiality in man's environment significantly, and have thereby generated new, non-natural needs. These technologies have created a range of objects (the communications satellite is one example) "imitating nothing and progressively added to by pure invention. And no less invented," Jonas adds, "are the ends which they serve. Power engineering and chemistry for the most part still added to the natural needs of man: for food, clothing, shelter, locomotion, and so forth. Communication engineering answers to the needs of information and control solely created by the civilization itself which made this technology possible, and once started, imperative." Jonas further calls our attention to the powerful hold which electrical-electronic technologies have on our imagination: "Whereas heat and steam [of mechanical technology] are familiar objects of sensuous experience, and their force is bodily displayed in nature, electricity," he writes, "is an abstract entity, disembodied, immaterial, unseen; and to all practical intents, viz., as a manipulable force, it is entirely an artificial creation of man."[12] Just as the technology of electronics transfigures the material world, so the metaphor of the invisible transforms our way of interpreting the world.

While it is generally agreed that the dominant image of the industrial age is mechanical and that the literary response to the machine is overwhelmingly negative, we have done less research on the response of the arts and literature to the technologies which have come after. We should assume that since technology itself has changed, we might find a different posture toward technology in the literature of the postmodern, postmechanical age and different metaphors to express that relationship. Mumford believes that mechanical technology is de-humanizing man, that art is opposed to technics. But in recent years we have witnessed a strong art *and* technology movement, of which John Cage is the exemplar, that questions and reverses Mumford's notions of art and technics and the roles they have to play in helping us achieve the good life. In recent years art has investigated, and paralleled, changes in technology. In

[12]Hans Jonas, p. 79, p. 78. Mumford was also aware of the impact of electrical technology on our daily lives. As early as *Technics and Civilization* (New York: Harcourt, Brace & World, 1963), he identified what he calls the neotechnic period (as opposed to the eotechnic or paleotechnic periods) with electricity and mass communications. He asserted that "lightness and compactness are the emergent qualities of the neotechnic ear" (p. 231) and enthusiastically noted that with electronic communications we are "now on the point of returning, with the aid of mechanical devices, to that instantaneous reaction of person to person with which it began . . ." (p. 239). In the thirties, his vision, although cautious, was radiant: "The dark blind world of the machine, the miner's world, began to disappear: heat, light, electricity, and finally matter were all manifestations of energy, and as one pursued the analysis of matter further the old solids became more and more tenuous, until finally they were identified with electric charges: the ultimate building stones of modern physics, as the atom was of the older physical theories." (p. 246). Later, as we see in *Art and Technics* and other writings, his prophecies darkened.

many quarters, the representation and generation of systems have succeeded the iconography of the machine. We find a de-materialization or ephemeralization of the art object, an emphasis on art ideas which are to be circulated as information rather than art objects which are to be tastefully consumed by connoisseurs. As Jack Burnham has observed, "by the mid-1960s a division had developed between the earlier 'machine art' and what could be defined as 'systems and information technology. . . . The definitive boundary line between the old and new technologies probably came with the New York Museum of Modern Art's 1968 exhibition, 'The Machine as Seen at the End of the Mechanical Age.'"[13] A spirit of apparent collaboration, not conflict, between artists and engineers, art and technics, has been on the rise. In 1966, for example, the engineer Billy Kluver organized "Nine Evenings: Theatre and Engineering" with John Cage and Robert Rauschenberg. In 1970 the Jewish Museum in New York mounted an exhibit based on computer technology called "Software." During this period the most ambitious project was Maurice Tuchman's five-year effort which paired twenty-two artists with corporations for the "Art and Technology" show at the Los Angeles County Museum (1967–71).

The art and technology movement rejects Mumford's idea of art as the expression of the personality, the exploration of the inner life. In fact, Cage believes that the imposition of the ego on the world is part of the problem, not part of the solution. Whereas Mumford would cast off the repression of the machine, Cage embraces the technics of electricity as a liberating force and adopts an impersonal system for generating his art. For Cage, technics moves toward the organic, art, toward the mechanical; technics gravitates toward the spiritual, art, toward the objective, or at least, the absence of the intensely expressive. If for Mumford, the values associated with technics are impersonality, regularity, efficiency, and uniformity, for Cage the values are heterogeneity, randomness, and plenitude. These are also the values he associates with art. In Cage's work, technics and art are not in conflict. Rather they are in the process of becoming more and more indistinguishable from one another. Thus Cage departs from the common Western definition of technology as a system of organized knowledge which has a merely instrumental, or practical, or

[13]See Jack Burnham's essay in this volume. For background on the art and technology movement, see also Jack Burnham, *Beyond Modern Sculpture: The Effects of Science and Technology on the Sculpture of this Century* (New York: George Braziller, 1968), Douglas M. Davis, *Art and Future: A History/Prophecy of the Collaboration Between Science, Technology, and Art* (London: Thames and Hudson, 1973), Lucy Lippard, *Six Years: The Dematerialization of the Art Object* (New York: Praeger, 1973), Jean Clarence Lambert, *Dépassement de l'art?* (Paris: Editions Anthropos, 1974), Pontus Hultén, *The Machine as Seen at the End of the Mechanical Age* (New York: Museum of Modern Art, 1968), *Pavilion*, ed. Billy Kluver (New York: Dutton, 1972), and Jonathan Benthall, *Science and Technology in Art Today* (New York: Praeger, 1972).

material purpose. Technology, according to *Webster's Seventh*, is "a technical means of achieving a practical purpose." More recent definitions of technology clearly specify that technology is not to be narrowly equated with machines or tools. As Bernard Gendron writes: "*A technology is any systematized practical knowledge, based on experimentation and/or scientific theory, which enhances the capacity of society to produce goods and services, and which is embodied in productive skills, organization, or machinery.*"[14] But even in this definition, the emphasis on practical purposes remains. For Cage, however, the technological can be more than merely instrumental. It can indeed be spiritual. His view of technology is zenlike: it is both intensely practical and mystical. Technology, he believes, can help us open our eyes and ears to the life we are living, "to the greatest mystery as it is daily and hourly performed."[15]

Just as Zen masters pose riddles which cannot be solved by logic or rationality (and Cage delights in these *koans*), so his work raises questions which cannot be neatly answered. In his texts, there seems to be an almost schizophrenic split between content and form, between responsible social commentary and aesthetics, and these two aspects of his work seem to imply different attitudes toward technology. Are they contradictory? Are they paradoxical? Depending on one's perspective (is it content? form? how can we separate the two?), we may conclude that Cage's ideas are either naive or wise. But Cage's work, as we know, eschews perspective. We come up against a transparent wall. We are left with questions, or with answers which must be questioned. As readers, we may be both baffled and intrigued, like the disciples of a Zen master.

II. How to Improve the World

The Aesthetics of the Text. The first installment of Cage's *Diary: How to Improve the World (You Will Only Make Matters Worse)* appeared in 1967. Six other sections have since appeared, with three others planned.[16] As the title indicates, the *Diary* presents a vague plan (or more accurately, a collection of attitudes and ideas) for the improvement of global social and economic conditions. The *Diary* thus falls into that most American of genres: the how-to-do-it-yourself manual. But, we will also be reminded that we will only make matters worse. Given the subject matter, we would not expect Cage's *Diary* to be a confessional account of an inner life, and it is not. Given the traditional form of the diary, we might expect

---

[14]Gendron, p. 23.

[15]D.T. Suzuki, *An Introduction to Zen Buddhism* (New York: Grove Press, 1964), p. 45.

[16]*Diary, A Year From Monday*, pp. 3–20, 52–69, 145–62, and *Continued 1968 (Revised), Continued 1969, Continued 1970–71, Continued 1971–72, M: Writing '67–'72* (Middletown, Conn.: Wesleyan Univ. Press, 1973), pp. 3–25, 96–116, 193–217.

this diary to be disjointed, non-continuous, and fragmented, and as we have already seen, it is, with a vengeance: Cage's *Diary* is composed of single words, personal anecdotes and amusing stories, brief quotations and references by proper name to over one hundred and fifty intellectuals, artists, and friends, and glances at topics ranging from the energy crisis and the absurdities of the American medical profession to dietary advice, mushrooms, and the Vietnam war. It is a bizarre, at times even scatological mixture of the domestic and abstract, a miscellany which is carefully compressed and yet relaxed and spontaneous. The *Diary* is what we could call concept writing, a telegraphic mine of bibliographical leads for the committed and curious in this time of information explosion. It directs us to McLuhan, Fuller, Illich, Mao, Mead, and Schumacher, to Kierkegaard, Thoreau, Basho, and Duchamp. Like the news, it comes to us in the form of a series of independent events which are not related to causal sequences. Like the statements of Zen masters, the parts of the *Diary* are presentations, not explanations or interpretations. The *Diary* is, by Cage's own description, a mosaic, as the following illustrates:

LXXIV.
Ephemeralization.   Away from the earth
            into the air.   Or: "on earth as it is
in heaven."   More with less: van der Rohe
            (aesthetics): Fuller (society of
world men).   Nourishment via odors, life
            maintained by inhalation: Auguste
Comte (Systeme de Politique Positive,
            second volume.   *Individuality.*   Out
            of the darkness of psychoanalysis
into sunny behavioral psychology (people
            picking up their couches and
walking).   **U.S. highway diner: now that
            I haven't eaten the potatoes, they
will throw them away (they should have
            been thrown away before being served).**
Rich, we become richer.   No way once it
begins to impede accumulation.
            Universe.   They've put the cart
before the horse: they're better about
publicize.[17]

The internal fragmentation of the text may lead us to conclude that it was composed haphazardly. This chaos is an illusion, however, for in fact the

[17]*Diary, A Year From Monday,* pp. 152–53.

parameters of the text are painstakingly determined by chance operations. The form of the *Diary* is generated by strict rules regarding the
length of the entries and their manner of presentation typographically. It
is governed, in other words, by a kind of indifferent organizational
machine. Cage's instructions to himself and Clark Coolidge, the printer/
publisher, are these:

> each day, I determined by chance operations how many parts of the mosaic I
> would write and how many words there would be in each. The number of
> words per day was to equal, or by the last statement written, to exceed one
> hundred words.
>
> Since Coolidge's magazine [*Joglars*, in which the first installment was
> printed] was printed by photo-offset from typescripts, I used an IBM Selectric
> typewriter to print my text. I used twelve different typefaces, letting chance
> operations determine which face would be used for which statement. So, too,
> the left marginations were determined, the right marginations being the
> result of not hyphenating words and at the same time keeping the number of
> characters per line forty-three or less.[18]

Cage uses chance operations in his musical and verbal compositions in
order to avoid imposing personal taste, memory, or the ego on the work
in a pre-determined way. He devises a system for self-effacement which
is both a means of abandoning control and a means of controlling himself
rigorously. There is nothing capricious about his work. His desire is to
frame a field in which a multiplicity of events is possible. The first
requirement, as in Zen, is to let go, to loosen the hold of the mind on the
world. As Cage says in his "Lecture on Nothing" in *Silence*, in which the
white spaces stand for silence:

				But
now		there are silences		and the
words	make	help make		the
silences		.		

                                        I have nothing to say
        and I am saying it   and that is
    poetry   as I need it   .
        .   This space of time   is organized
                        We need not fear these   silences,—
                        ₥

    we may love them .
                                            This is a composed
    talk   ,   for I am making it
        just as I make   a piece of music.   It is like a glass
            of milk   .   We need the   glass

[18]*Diary, A Year From Monday*, p. 3.

and we need the	milk	.	Or again	it is like an
empty glass		into which		at any
moment	anything		may be poured	
.	As we go along			

an i—dea may occur in this talk   .

Only if a structure is shaped scrupulously and indifferently is true novelty possible, only then may an idea "occur." Nothing can happen if everything is pre-determined. With Cage, both game and play, two forms of activity difficult to reconcile, are contingent upon one another, as are determinism and chance. Is this paradoxical? contradictory? The system is impersonal, but the resulting text is serenely personal. The rules for generating the text of the *Diary* are rigid, the moves require a stylized, meticulous labor, but the resulting space on the page is thus prepared for play: a gratuitous activity with no instrumental goal in mind: anything can happen. Thus in the *Diary* we encounter no dominating narrative, no linear thread which connects the fragments. In the *Diary*, the events of the day, the thoughts of the evening, are allowed to surface within the bounds of structural constraints. We have, as Cage puts it, not a *series* of components, but a series of *components*. He believes that freedom inheres in the space between them, but beyond this, he also believes that they form a system by a kind of relaxed association of ideas, much as do conversations which ramble pleasantly between friends. We discover this system, he believes, we do not create it. All things are related in this universe, if we allow them to appear to us: this is his faith in the sweet coherence of the world. The logic is part Eastern, described by Lama Anagarika Govinda this way: "If we consider, instead of a sequence, the simultaneity of certain seemingly unconnected phenomena, we shall often be able to observe a parallelism, a coincidence of certain qualities, not causally or temporally conditioned, but rather giving the impression of a cross section of an organically connected whole."[19] Cage would object to the word "whole," but would agree with the rest. In the following two sections from the first installment of the *Diary*, we see how Cage's invention of an impersonal system (the complicated strategy of chance operations, a game of sorts) has generated a specific structure (the particular parameters of the text) whose playful content, he believes, can help us open our eyes and ears to another system: the world.

---

[19]Lama Angarika Govinda, "Logic and Symbol in the Multi-Dimension Conception of the Universe," *Main Currents in Modern Thought*, 25 (Jan.-Feb. 1969), 61.

III.   AS McLUHAN SAYS,
            EVERYTHING HAPPENS AT ONCE.   IMAGE IS
NO LONGER STREAM FALLING OVER ROCKS,
            GETTING FROM ORIGINAL TO FINAL PLACE;
IT'S AS TENNEY EXPLAINED: A VIBRATING
                    COMPLEX, ANY ADDITION OR SUBTRACTION
OF COMPONENT(S), REGARDLESS OF APPARENT
        POSITION(S) IN THE TOTAL SYSTEM,
        PRODUCING ALTERATION, A DIFFERENT MUSIC.
        FULLER: AS LONG AS ONE HUMAN BEING IS
                    HUNGRY, THE ENTIRE HUMAN RACE IS
HUNGRY.   City planning's obsolete.   What's
                    needed is global planning so Earth
        may stop stepping like octopus on its
own feet.   Buckminster Fuller uses his
        head: comprehensive design science;
inventory of world resources.   Conversion:
            the mind turns around, no longer
                    facing in its direction.   Utopia?
                    Self-knowledge.   Some will make it,
    with or without LSD.   The others: Pray
        for acts of God, crises, power
        failures, no water to drink.   *IV.   We
see symmetrically: canoe on northern
                Canadian lake, stars in midnight sky*
*repeated in water, forested shores
    precisely mirrored.   Our hearing's
asymmetrical: noticed sounds surprise us,
echoes of shouts we make transform our
voices, straight line of sound from us to
                shore's followed by echo's slithering
        around the lake's perimeter.   When I
                said, "Fifty-five global services,"*
*California Bell Telephone man replied
        (September '65), "It's now sixty-one."*
The seasons (creation, preservation,
                destruction, quiescence): this was
                experience and resultant idea (no
longer is: he flies to Rio).   What shall
we wear as we travel about?   A summer suit
    with or without long underwear?   What
                about Stein's idea: People are the way
their land and air is?

[20]*Diary, A Year From Monday*, pp. 4—5.

From this excerpt, we see that Cage is committed to the idea of systemic change rather than linear causation ("IMAGE IS NO LONGER STREAM FALLING OVER ROCKS, GETTING FROM ORIGINAL TO FINAL PLACE"): to grasp this, we need to open our ears for *"our hearing's asymmetrical."* Further, Cage agrees with Buckminster Fuller that we must think in terms of a global economic system, not cities or nations, for communications technology has spanned the planet (*"When I said, 'Fifty-five global services,' California Bell Telephone man replied [September '65], 'It's now sixty-one.'"*). Moreover, he concludes from the changes brought about by technology that our experience of time and space has radically changed. No longer do we interpret our experiences on the basis of the eternal round of seasons in a fixed spot; both the seasons and space have become discontinuous:

> The seasons (creation, preservation,
> destruction, quiescence: this was
> experience and resultant idea (no
> longer is: he flies to Rio).   What shall
> we wear as we travel about?   A summer suit
> with or without long underwear?   What
> about Stein's idea: People are the way
> their land and air is?

A Technology of Liberation: Electronics as Means and Metaphor. In terms of aesthetics, then, Cage relinquishes a large measure of control, giving it over to a system which he has himself devised and which I have described as an indifferent organizational machine. The analogy suggests the extent to which Cage, both a mechanist and a mystic, places his trust in technology as a benevolent force which, once set in motion and allowed to continue undisturbed, will improve the world. I have also stressed the difference between the programmatic structure of the *Diary* (the system for generating the form of the text) and the open, anarchic nature of its internal strategy, or method, because these distinctions correspond to Cage's vision of the uniform structure of social and economic organization on the planetary level and the multiplicity which is possible on the local level. For Cage, the work of art provides us with an analogue for the best possible organization of culture. Likewise, his pleasure in the *Diary* with the variety of elements offered by the IBM Selectric is analogous to his conviction that technology can find happy and diversified solutions to many of our current social problems.

"Only chance to make the world a success for humanity," he writes in characteristic *Diary* shorthand, "lies in technology, grand possibility technology provides to do more with less, and indiscriminately for everyone."[21] The goal is traditionally American, or so it at first appears:

> (What we want is very
> little, nothing, so to speak.   We just
> want those things that have so often
> been promised or stated: Liberty,
> Equality, Fraternity; Freedom of this and
> that.)[22]

Liberty, Equality, Fraternity, Freedom: to reach this ideal state of affairs we must begin, Cage advises, by satisfying basic human needs. His list of these needs is revealing: "water, food, shelter, clothing, electricity, audio-visual communication, transportation."[23] According to Hans Jonas, some of these needs are "natural," but the need for access to mass communications networks and information produced by electrical-electronic technologies is artificial, new. According to Cage, all of these needs are basic and can be fulfilled by technology. Cage envisages, in other words, a technology of liberation.

Like Fuller, whom he deeply admires, Cage assumes that we can satisfy all of our material needs through technology. The fundamental problem is no longer that of scarcity. With comprehensive design science, we can produce enough food and shelter to support our global population comfortably. "Begin again," writes Cage, "assuming abundance."[24] And like Marshall McLuhan, the other major intellectual spirit behind the *Diary*, Cage realizes that we are living in an era of electronic communication. In the *Diary* we find a very Cageian blending of these two post-industrial prophets: the Spaceship Earth is a congenial global village. This mixture of Fuller and McLuhan lifts Cage's vision of a utopian world beyond the crass materiality of the American dream. Cage would abolish the sacred American institution of private property, including intellectual property, and substitute *use* as the yardstick for temporary possession (he suggests, for example, that we occupy houses as we need them, just as we do phone booths). For Cage, what is essential to the development of human potential is not material abundance, the American cornucopia of manufactured plenty. The style of this work, in fact, suggests the very opposite. Marked by a sparseness and delicacy antithetical to the weight of goods, the language of the *Diary* has an invisible, light, non-object quality which we do not find in some other postmodern writing—concrete poetry, for example, or Raymond Federman's *Double or Nothing*,

[21]*Diary, M,* p. 102.
[22]*Diary, A Year From Monday,* p. 155.
[23]*Diary, A Year From Monday,* p. 15.
[24]*Diary, A Year From Monday,* p. 157.

or Michel Butor's *Mobile*. In place of material abundance, Cage substitutes an *abundance of information*.

This is what Cage means when he says that it is electronics (particularly in the form of telecommunications and the computer) which can alter global society. To his mind, the implications of this post-industrial technology are democratic: electronic communication will transform hierarchies of power into horizontal networks. For example, he suggests a global lecture series on war. He endorses Fuller's notion of electronic democracy (voting by television):

> Summit lecture series on War: not to be
> given in one city, but via a global
> Telstar-like facility, each receiving set
> throughout the world equipped with a
> device permitting hearing no matter
> what speech in one's own tongue.[25]

He is charmed by the fact that the Russians have learned to induce sleep electronically, thereby cutting the need for eight hours of sleep to thirty minutes. He proposes an investigation into electronic clothing. He believes that instant communications will lead to one language and one culture, and this will be beneficial. But more important than these fanciful, specific proposals is Cage's conviction that technological developments of the electronic era will be accompanied by a positive change in human values. The growth of electronics will bring about an increase in personal and global flexibility. As he puts it,

> When our time was given to physical labor, we needed a stiff upper lip and backbone. Now that we're changing our minds, intent on things invisible, inaudible, we have other spineless virtues: flexibility, fluency. Dreams, daily events, everything gets to and through us.[26]

Thus for Cage electronics is much more than a concrete phenomenon. It is also a metaphor for what is invisible and thus spiritual. In the post-mechanical era, in the work of John Cage, the metaphor of the dehumanizing machine is displaced by images of electronics as a spiritual force.[27] Through electronic technology, Cage believes that we are extending our mind(s) into a world which has been separate from man for a long time, the world of nature. In a passage which recalls the mysticism of Teilhard de Chardin and the theory of Marshall McLuhan, Cage writes: "we have only one mind."[28] His position is that of an idealist. It is the wire*less*

[25]*Diary, A Year From Monday*, p. 152.
[26]"Diary: Audience 1966," *A Year From Monday*, p. 51.
[27]See Ihab Hassan, *Paracriticisms: Seven Speculations of the Times* (Urbana: Univ. of Illinois Press, 1975), pp. 121–47, for speculations on the gnosticism of the postmodern imagination.
[28]*Diary, A Year From Monday*, p. 158.

technology which can turn man toward his original harmony with nature:

> We need for instance an
> utterly wireless technology.   Just as
> Fuller domes (dome within dome,
> translucent, plants between) will give
> impression of living in no home at
> all (outdoors), so all technology
> must move toward way things were
> before man began changing them:
> identification with nature in her manner
> of operation, complete mystery.[29]

As Cage poignantly puts it, we must "treat redwoods, for instance, as entities that have at least a chance to win."[30] Extended, this benevolent attitude toward nature becomes more than mere respect. One arrives at identity: man and nature are one. In one of the few lyrical passages in the *Diary*, Cage describes the landscape he sees while driving toward Chicago, and concludes parenthetically that "We, too, are trees."[31] Everything is necessarily connected in one global system which is both material and spiritual. And it is the computer, he says, which is "bringing about a situation that's like the invention of harmony."[32] Thus Cage proposes a theory and practice of global systemics, an ecology of man, culture, and nature.

On the structural level, Cage's model for a utopian planetary culture based on abundant information is the global utility. The utility, Cage supposes, has an organization which is almost invisible. He believes that like any other machine, the utility possesses the virtue of being disinterested in power. Thus Cage does not believe that technology has gone out of control, although he is a technological determinist of sorts. Nor is the individual to blame for the state of man's affairs. Cage optimistically believes he will thrive if only given healthy conditions. (What is needed, Cage quips, is electroanalysis, not psychoanalysis.) Rather it is the economic-political system which inhibits the productive use of technology (Cage is not precise, however, about what he means by this). How will change be effected? Cage regards the growth of global services as auguring the withering away of the state. We should note here that this social tenet is, like others, analogous to his aesthetic theory. For Cage, music is contemporary if it can be successfully interrupted (that is, joined) by environmental sound. Likewise, he recognizes a sculpture or painting as contemporary if it is able to incorporate environmental

[29] *Diary, A Year From Monday*, p. 18.
[30] *Diary, A Year From Monday*, p. 11.
[31] *Diary, M*, p. 8.
[32] *Diary, M*, p. 22.

light and shadows as a legitimate part of the piece. He writes:

> Whether or not a painting or sculpture lacks a center of interest may be determined by observing whether or not it is destroyed by effects of shadows. (Intrusions of the environment are effects of time. But they are welcomed by a painting which makes no attempt to focus the observer's attention.) Observe also those works of painting, sculpture, and architecture which employing transparent materials, become inseparable from their changing environments.[33]

Similarly, Cage believes that government is "contemporary if its activities aren't interrupted by the action of technology."[34]

If Cage welcomes the standardization of technology on the global level in the form of the global utility, it would follow (it would seem) that he more or less accepts longterm economic planning and coordination of production and distribution of resources. This problem does not receive much attention in his work, although he does offer specific, minimal proposals to this end, such as standardizing garbage cans and establishing "a global voltage, a single design for plugs and jacks."[35] But perhaps more important are the implications we can draw from his aesthetic theory. As we have already seen, his model for global organization corresponds to his system for generating the parameters of the *Diary*. And just as chance operations encourage heterogeneity and refuse the domination of narrative in the text of the *Diary*, so the world utility, generating an abundance of information and a multitude of channels for communication, will increase, Cage asserts, the possibilities for diversity. He writes, referring to global voltage, "vary not the connecting means but the things to be connected."[36] Standardization of structure will not necessarily imply homogeneity of behavior. We need not fear a mass culture peopled with automatons and punctuated with identical cultural artifacts. As the globe becomes organized, Cage maintains, life will remain pleasantly disorganized. In other words, according to Cage, the dreams of both Fuller and Thoreau are mutually compatible and in fact interdependent. The former promises *equality* in a *mass society* while the latter promises *freedom* and *diversity* within a *global village*. Cage thus combines in one vision the two dominant scenarios projected for postindustrial society—on the one hand, an organized planetary culture, and on the other, clusters of tribal, agricultural communities which are de-centralized.[37] The rationalization of resources

[33]*A Year From Monday*, p. 31.
[34]*Diary, M*, p. 18.
[35]*Diary, A Year From Monday*, p. 55.
[36]*Diary, A Year From Monday*, p. 55.
[37]Michael Marien, "The Two Visions of Post-Industrial Society," *Futures*, 9 (October 1977), 431–51.

can go hand in hand with the liberation of human sensitivity and sensibility. What Cage seeks—and he borrows the terminology of Richard Kostelanetz to describe it—is "technoanarchism."[38] As we read in Cage's Afterword to *A Year From Monday:*

> now that we hunt for signs of practical global anarchy, such signs appear wherever we look. We see a current use for art: giving instances of society suitable for social imitation—suitable because they show ways many centers can interpenetrate without obstructing one another, ways people can do things without being told or telling others what to do. We look forward to "an environment," N.O. Brown, "that works so well that we can run wild in it." We are evidently going to extremes: to the "very large scale" ("the great impersonality of the world of mass production") and to the "very small scale" ("the possibility of intense personalism")—Edgar Kaufman, Jr., *The Architectural Forum*, September 1966. We anticipate "the dwindling," the eventual "loss of the middle scale" (*Verlust der Mitte*—J. Strygowsky): bureaucracies associated with power and profit. The information with which these bureaucracies now deal in a stalemated fashion (C. Wright Mills: *The Power Elite*) may be relegated to our computers, our unimaginative rationality put down under as in a sewer—N.O. Brown—so that "the reign of poetry"—Thoreau, Brown, and the rest of us—may at long last commence.[39]

III. The Paradoxes of the Text.

Cage's music has been greeted with incredulity and hostility. So too his optimistic vision of our future enrages many of his readers. Cage simply cannot be taken seriously, they say, he is too naive, he doesn't understand the political process, he has no grasp of the profound irrationality of human nature. His imperturbability, while charming, can be exasperating. In part, it is because he is committed philosophically to accepting what we in the West would define as contradictions. He refuses the Western habit of thinking in terms of bipolarities. Searching for the common denominator between Mao and Fuller and encountering seemingly irreconcilable differences between the two, for example, he has decided to listen to both: "For instance, Fuller's advice, 'Don't change man; change environment,' and Mao's directive: 'Remould people to their very souls; revolutionize their thinking.'"[40] Such directives are no more opposed to one another, he believes, than are mountains opposed to spring weather. The premise of his work is that we suspend argument and listen. On the one hand, this is good advice: to open up our minds.

[38]*Diary, M*, p. 113.
[39]*A Year From Monday*, pp. 165–66.
[40]Foreword, *M*.

The *Diary* can be seen as a Cageian bibliography to postindustrial prophets, a speculative card catalogue, a friendly postmodern reading list. It invites us to make our connections among the constellations of information events, to do our homework, to read wisely, and to make up our own minds.

On the other hand, the form of the *Diary* does pose a very thorny problem. In his musical compositions, Cage gives up a measure of control, rejecting point of view for both himself and his listeners. Everyone, he tells us, is in the best seat. This aesthetic democracy of sound, this de-centering of music, is both engaging and significant. Playing with words, however, is different from playing with sound. There is no democracy in affairs of the mind. To be a responsible thinker about culture on this level requires that one adopt a point of view. Thus, while we may find that the form of the *Diary* is fruitful if it is conceived as a bibliography, we may also find it intellectually fractured if understood as an argument. This is also a paradox, or at least a contradiction. It would appear at first that the *Diary* is a "writerly," not a "readerly" text, to use Roland Barthes' terms. The readerly text, or classical text, puts the reader in the position of being a consumer rather than a producer of a text. "Instead of functioning himself," Barthes writes, "instead of gaining access to the magic of the signifier, to the pleasure of writing, he is left with no more than the poor freedom either to accept or reject the text: reading is nothing more than a *referendum*." Like the "writerly" text, Cage's *Diary* asks us to make our own text, it is plural. His text cannot be consumed or remembered. "The writerly [text]," explains Barthes, "is the novelistic without the novel, poetry without the poem, the essay without the dissertation, writing without style, production without product, structuration without structure."[41] Yet Cage's text is, covertly, also a "readerly" text. For while the method of composition gives us the impression that Cage is not taking a position, the cumulative weight of the fragments constitutes a point of view. A utopian statement about electronic technology does emerge, but not as a result of sustained intellectual argument. In front of this text we are stymied. We are not openly presented with an argument, we chance instead upon references to other people's work. Thus to argue rationally against the content of the *Diary* would require us to take on McLuhan, Fuller, Mao, Thoreau, Schumacher, and all the others, and what a mind-boggling intellectual olympics that would be. At this point we may understandably feel like the Zen disciple who was asked to describe the sound of one hand clapping. We may find ourselves wanting both to accept and reject the implications of the *Diary*. In order to untangle this riddle, if that is possible, first we will

[41]Roland Barthes, *S/Z*, trans. Richard Miller (New York: Hill and Wang, 1974), pp. 4–5.

look at the *Diary* as an electrical utopia and then turn to its form, while acknowledging that to split the *Diary* into two is to violate the very spirit of the text.

The Electrical Utopia. In their long and persuasive essay on "The Mythos of the Electronic Revolution," James Carey and John Quirk show that the rhetoric of the electrical sublime, as they call it, has hardy roots in American culture extending back to the mid-nineteenth century.[42] Cage's ideas, we learn, fall into a genre which celebrates electricity as a magical force, a panacea which will bring about the American ideals of abundance, democracy, community, and harmony among man, nature, and technology. Carey and Quirk point to numerous instances of what they judge is innocent and naive rhetoric, including the nineteenth-century economist Henry Charles Carey's *The Unity of Law*, William Dean Howell's *A Traveler From Altruria*, Lewis Mumfords's *Technics and Civilization*, David Lilienthal's dedicatory address to the Tennessee Valley Authority, Marshall McLuhan's work, and pronouncements of Zbigniew Brzezinski. This rhetoric, Carey and Quirk conclude, prevents us from analyzing or understanding the real effect of electronic technology. Worse, it serves the interests of the electric light and power companies. Carey and Quirk locate the most trenchant theoretical criticism of the electronic utopia in the work of the Canadian political economist Harold Innis who argues that the technology of electricity, particularly in the form of mass communications, necessarily promotes centralization and the expansion of an imperialist technocracy. They summarize Innis' argument this way:

> Innis placed the tragedy of modern culture in America and Europe upon the intrinsic tendencies of both printing press and electronic media to reduce space and time to the service of a calculus of commercialism and expansionism.
>
> . . . .
>
> Innis assessed the importance of historical and geographical factors and their relation to the means of communication and transportation. He developed from those assessments the theory that the ways in which communication and transportation systems structure (or "bias") relations of time and space were at the base of social institutions. Innis divided communication and social control into two major types. Space-binding media, such as print and electricity, were connected with expansion and control over territory and favored the establishment of commercialism, empire and eventually technocracy. On the other hand, time-binding media, such as manuscript and human speech, favored relatively close communities, metaphysical speculation, and traditional authority.

[42]James W. Carey and John J. Quirk, "The Mythos of the Electronic Revolution," *The American Scholar*, 39 (Spring 1970), 219–41 and (Summer 1970), 395–424.

Carey and Quirk then show that recent developments in American history demonstrate that electronic technology has indeed increased centralization, pollution, and conflict. They end their essay with an appeal to intellectuals to de-mystify this faith in electronic technology. "Electronics," they write, "is neither the arrival of apocalypse nor the dispensation of grace. Technology is technology; it is a means for communication and transportation." Further, "the demise of culture could only be offset by deliberately reducing the influence of modern technics and by cultivation of the realms of art, ethics and politics. This requires action to counter and direct rather than to disguise the bias of the electronic revolution: it means cultural and qualitative checks rather than more quantitative definitions of the quality of life; it requires defusing the humanistic from the technological instead of offering a contradictory image of humanized technology."[43]

I am in agreement with Carey and Quirk: the rhetoric of the electrical sublime should be seriously questioned. They persuade me, in fact, that it should be dismissed. Moreover, we can make other objections to Cage's notions, all of them sound. We might protest that Cage makes the mistake of equating information with meaning. We might agree with the biologist C.W. Waddington that Cage's interest in social *dis*organization cannot satisfy the human desire for the social bond and, worse, masks a traditional American belief in individualism. As Waddington has said:

> I think that this anarchy that John Cage is preaching is really a sort of upside-down version of the American dream—which has turned out to be, in practice, a dream of the freedom and self-sufficiency of the individual to experience rather than to acquire. But, moving among all these public utilities—cars you pick up and drop as and when required, dwellings you go into and quit as you might a telephone booth—the person would be free but remain an isolated unit; none of these things he uses would belong to him, or connect him to anyone else, they would be socially as neutral as the air we breathe or the piped water we drink. I think that biology shows that human life needs more social integration than that.[44]

And we might object that Cage's quixotic quietism is naive: he does not grapple with questions of power, he sees them as irrelevant. As he has written, "my notion of how to proceed in a society to bring about change is not to protest the thing which is evil, but rather to let it die its own death."[45]

---

[43]Carey and Quirk, pp. 238–39, 422, 423.
[44]*Biology and the History of the Future: An IUBS/UNESCO Symposium* (Edinburgh: Edinburgh Univ. Press, 1972), p. 30.
[45]*Biology and the History of the Future*, p. 35.

A Systemic Approach. If one were to map American attitudes toward technology, Cage would be characterized best as a happy technophile. In terms of content, this is clearly a naive stance. In terms of method, however, Cage's contribution to our thinking about technological control is sophisticated. Although Cage cheerfully embraces an electronic technology of liberation, he can in no way be labelled a technocrat. As many have recognized, Cage's writing does not so much tell us what to think as how to think in a new way. How to improve the world? Cage understands that there is every likelihood that rational, conscious control will only make matters worse. His strength is that he understands the problematic inherent in intervening consciously in a complex system, in attempting to dominate the future. His method is appropriately cautionary. He shares Gregory Bateson's conviction that the role of art in technological culture is to act as a corrective to our too-purposive view of the world and consciousness. Art calls attention to a systemic view of life. This means that the wisdom of art is humility: the knowledge that the human is only part of a larger system and that the part can never ultimately control the whole.[46] Cage writes:

> man-made structures themselves . . . must give way if those beings they were designed to control, whether people, animals, plants, sounds, or words, are to continue on earth to breathe and be.[47]

Cage's antidote to technological determinism/autonomy is indeterminacy. His genius is that he has invented a model of a system which is put into play by the self and is at the same time self-propelling, a system which is determined and yet allows for the unexpected. Cage's method of constructing a text through chance operations is a unique blending of East and West, of Zen and a delight in an impersonal order achieved through a mechanism. Art and technics merge. One balances the other. Together they encourage difference, variety, multiplicity. Cage's aesthetics are analogous to a basic biological principle: increasing heterogeneity promotes the health of a culture. Thus the *Diary* is a non-narrative utopian parable, an exemplum of how not to impose oneself. Unlike other contemporary utopias (whether they be traditional narratives, like Ursula LeGuin's *The Dispossessed,* or non-fictional products of systems design, like the projections from the Club of Rome), Cage's *Diary* does not construct a closed system:

[46]Gregory Bateson, *Steps to an Ecology of Mind* (San Francisco: Chandler Publishing Co., 1972).
[47]Foreword, *M.*

> Begin again, assuming abundance,
> unemployment, a field situation,
> multiplicity, unpredictability,
> immediacy, the possibility of
> participation.

Neither a blueprint for society nor a mere paper carton of bibliographical information, the *Diary* is, to use Darko Suvin's phrase, "a methodological organ for the *New* . . . a heuristic device for perfectibility, an epistemological and not an ontological entity."[48] It is an invitation to listen, to experience, to do. As Cage has said, "art is a means of self-alteration, and what alters is the *mind*, and mind is in the world and is a social fact. . . . We will change *beautifully* if we accept uncertainties of change; and this should affect any planning. This is a *value*."[49]

---

[48]Darko Suvin, "Defining the Literary Genre of Utopia," *Studies in the Literary Imagination*, 6, No. 2 (1973), 135.
[49]*Biology and the History of the Future*, p. 46.

MUSIC, VOICE, WAVES*

DANIEL CHARLES

I. Music

The aim of Murray Schafer's "Soundscape Project" is to treat the entire world as a macrocosmic musical composition. As the author notes, such a program is based on John Cage's new definition of music:

> Music is sounds, sounds around us whether we're in or out of concert halls: cf. Thoreau.[1]

Before the sixties this definition of music would have been unimaginable. Only as recently as 1958 did the poetics of indeterminacy surface in a piece by Cage: his instructions for "Concert for Piano and Orchestra" specified that one of the possibilities for performing the composition was *non*-performance, which implies that we hear the piece even when it is not played, and that everything we do hear is a part of it, thus *all* of the sounds of the universe are welcome as music, whether they received a written invitation from Cage or not.

*I would like to thank M. Michel Sanouillet, who suggested the title of this paper and welcomed me so warmly at the May 1978 Conference on "Technologie et culture post-industrielle" in Nice.
[1]John Cage, quoted in R. Murray Schafer, *The New Soundscape* (Don Mills, Ontario: BMI Canada Limited, 1969), p. 1.

The problem of the composer is thus a problem of *ecology*. As Murray Schafer shows,[2] the fantastic explosion of percussion in the mainstream of Western music since the nineteenth century parallels the increase in non-pitched and a-rhythmic sounds. In addition, in reaction against serial composition, there has been an increase in the use of chance operations and aleatory techniques, which surrender the "rational" organizing of sounds in musical compositions to the "higher" laws of entropy. (Such compositions are in fact spatial-temporal frames, that is, concert halls, which have opened up their doors and windows to the domain of sound outside, as in Cage's 4'33", a "transparent" piece which refuses to impose itself on the environment: indeed, art and "life" are identical to one another.) In addition, the inclusion of any and all sounds from the environment in a piece of music—in composition via tape (*musique concrète*), for example, or the synthesizing of sounds (electronic music and the computer)—has worked to efface the composer: music is "alive" to the extent to which it participates in the *life of machines* rather than reflects the free will and taste of the artist. Thus today's orchestra is the entire sonic universe; today's musicians include everyone and *everything*, whether living or not, "sentient being or non-sentient being" (D.Z. Suzuki, cited by Cage), provided that their relationship to one another is characterized by "interpenetration AND non-obstruction."

The result: with this skewing, this displacing of social roles, the body also is redefined and given a new importance. According to Cage, to compose is one thing, to play is another, and to listen, still another. In experimental music, in which the composer cannot foresee what will happen, this separation of the very *gestes musicaux* explodes the chains of causality. To compose is to paint; to play is to give oneself up to a dance which bears no relationship to the sounds themselves; to listen is thus all that remains of what we have traditionally understood as music. But to listen radically, to listen for the first time, that is, contemporary listening—which involves equally the composer and the performer as well as the audience—has nothing to do with memory or the anticipation of the future. This kind of listening always returns us to the primacy of that which is *heard* over that which is *seen*, which anthropologists claim is characteristic of rural African culture, that is, oral culture.[3]

Listening is thus a function of a new priority conferred on the ear in a culture which favors the eye and the image. Listening is a means of touching at a distance—and it is variations of distance and therefore the differentiations in auditory nearness which raise a fascinating philosophical problem. For we need only remember that in the West the privacy of

[2]Murray Schafer, *The Tuning of the World* (New York/Toronto: Alfred A. Knopf/ McClelland and Stewart Ltd., 1977).
[3]Schafer, *The Tuning of the World*, p. 11.

one's inner life has always been linked to the turning inward of the ear. As Murry Schafer maintains, following Adorno, the ear, unlike the eye, cannot be closed at will and can be taken at any time for an erotic orifice: "Listening to beautiful sounds, for instance the sounds of music, is like the tongue of a lover in your ear."[4]

II. Voice

The $\begin{Bmatrix} \text{language} \\ \text{tongue} \end{Bmatrix}$ of your lover!

To say tongue is to say mouth, breath, voice. And to say voice is to say—at least traditionally in the West—subjectivity. Insofar as it is perceived *inwardly* and *outwardly* at the same time, the sound of "my" voice allows "me" to construct "myself" as a subject. It is not only into someone else's ear that "my" tongue slips—it finds its way first and foremost into my own. "I" listen to "my"self speak—and in the doubling of the "I" by the "my," or in the "self" of "myself" which "I" voice by and through "my" tongue, filling, form within as well as without, "my" ear, I situate "my"self as a SUBJECT IDENTICAL TO ITSELF. My voice is both the *message* (I listen to "my"self speaking) and the *medium* (listened to as music, it establishes ME as the self which I indeed am). My voice confirms me in my being; it assures my continued existence, it creates the ground of my subjectivity.

This is how Western philosophers, listening to themselves speak, invented subjectivity, logocentrism, the "full" word, presence. If we are to believe Derrida, Plato distrusted the written word even more than does McLuhan. To recuperate writing, or at least to redefine it as pre-phonetic, pre-phonological or pre-phonic, would mean going against the grain of Western philosophy. One would need to puzzle out a primordial trace, anterior to the voice and to any problematic of origin, leading to an arche-trace, a *gramma* present in every genetic *pro-gramme* before any *Logos* or any phoneme.

Elsewhere I have developed a theory of the voice and the oral tradition which permits us to construct a philosophy of subjectivity without involving ourselves in the question of writing, such as we find in the Derrida of the *Grammatologie*. Although I can do no more here than point the readers to that theory,[5] I would, nevertheless, like to quote, *en hommage à* Michel Benamou, a significant passage which critiques Derrida and skirts by a route different from mine, the aporias of the split between the voice and writing in order to establish beyond this split, an ethno-

---

[4]Schafer, *The Tuning of the World*, p. 12.
[5]Daniel Charles, *Le Temps de la voix* (Paris: Delarge, 1978).

poetics which is closely linked to music as I understand it. Benamou
begins by quoting Derrida in *L'Ecriture et la différence:*

> there exist two interpretations of interpretation, structure, sign and free
> play. The one seeks to decipher, dreams of deciphering a truth or origin
> that would escape play and the order of signs, and feels exiled in the
> necessity of interpretation. The other, no longer turning toward an
> origin, affirms free play and tries to go beyond man and humanism, the
> name of man being the name of this being who, throughout the history
> of metaphysics or onto-theology, i.e., his whole history, has been
> dreaming of full presence, reassuring foundation, origin and an end to
> play.

We are dealing here with two marginalities: the marginality of speech
recalling the oral tradition and its supposed mysteries of innocent presence,
and a marginality of writing found today in the poetics of free play and chance
operation: from Mallarmé to Oulipo, from John Cage to Jackson Mac Low.
Derrida's critique of ethnology raises a question about mantras, formulaic
poetry, and the respect for the oral tradition. But the question is not simply
whether tribal authority refers us to an original Logos as Law and Creation.
There is a large distinction to make between Greek Logos and African
Nommo, an instance of divinity shared by all who have the power of the
word. The question—surely not instantly soluble now—is how to recover the
word, in Dennis Tedlock's formulation, or in William Spanos':

> It seems to me that everything I've heard in this symposium [a
> symposium on Ethnopoetics, Center for Twentieth Century Studies,
> University of Wisconsin-Milwaukee, April 1975] so far hasn't addressed
> itself to this Heideggerian notion of speech (*Rede*) or has confused a
> logocentric notion of orality, which is grounded in an abiding presence
> or in a telos, an orality, that is to say, which begins meta-physically,
> begins, in other words, from the end, with an existential notion of orality
> that is committed to a process beginning not far from something but from
> nothing, from zero . . . . We've got to distinguish between these two
> kinds of orality, what I would call the spatial and temporal modes.

A cross-over has taken place for the American poet who, grounding his
secondary (re-learned) orality on the tribal communitas, is reaching back to
primitive graphism; beyond the barbarian order of writing, beyond the
civilized order of print, beyond justified margins and typographical conform-
ities, to a non-alignment of voice and writing essential to marginality:
recovering the voice *and* the sign. The new communitas of performance, as in
[David] Antin's semi-improvised chamber sessions and the dadaist act of
exploding the print culture now connect in the rhizome of ethnopoetics.[6]

---

[6]Michel Benamou, "Postface: In Praise of Marginality," in *Ethnopoetics*, ed. Michel
Benamou and Jerome Rothenberg (Boston: Alcheringa/Boston Univ. Press, 1976), p. 138.

III. Waves

As I have pointed out, experimental music, as defined by John Cage, leads us to a reconsideration of the function of the ear and thus of the role of the voice. I have stressed this point by referring to Michel Benamou's remarks on logocentrism in Derrida. If, as both William Spanos and Michel Benamou believe, a temporal orality exists, it corresponds quite closely to what I have called *"le temps de la voix"* and ethnopoetics and musicology are thus no longer estranged from each other. Both disciplines demand the voice be re-possessed not only as an interior phenomenon, but also as an exterior phenomenon—as in Murray Schafer's "Soundscape."

The voice is first a channel, a path, a course, a journey in a polyphonic forest; the voice defines a space or a region in terms of time well before it shapes interior space. To use the distinctions of Deleuze and Guattari in *Rhizome,* such a project is a kind of mapping, like the charts of Cage's scores, rather than a kind of tracing or like the process of transferring decals:

> Unlike a tracing, a map is always involved in trying to capture the real. . . . A map has many points of entry, while a tracing always returns us to the same thing. A map generates performances, while a tracing always refers to a "would-be authority."[7]

Without a doubt, although there are fundamental differences between a map and a tracing, there are certainly many points of similarity between them. Nevertheless, the rhizomatic nature of a map can never be reduced to the sedentary, stiff, rigid nature of a tracing: "Take psychoanalysis and linguistics, for example: the former has taken only photographs or tracings—nothing more—of the unconscious, the latter, photographs or tracings of language, with all the inaccuracies and slippages which that implies (it is not at all surprising that psychoanalysis has allied itself with linguistics)."[8] But neither psychoanalysis nor linguistics can take more than *flash photographs* of the voice. Music, on the other hand, modulates these flashes, as in the work of Demetrio Stratos, for example, where one, two, three basses sound in unison, of one accord, one voice which becomes manifold, a multidimensional, in-depth *frayage.* This is a rhizomatic voice (*"voix-rhize"*) which reproduces nothing, especially not the deadends and deadlocks of a STYLE.

If contemporary music consists in the proliferation of voices, we can say that it is wave-like rather than particle-like. Concerned with *process,* it goes beyond any notion of the production of objects, in the sense that

---

[7]Deleuze and Guattari, *Rhizome* (Paris: Ed. de Minuit, 1976), pp. 37–38.
[8]Deleuze and Guattari, p. 41.

music has traditionally always devoted itself only to the ready-made, skillfully made to measure, to be sure, but ready-made nonetheless, that is, *"com-positio,"* a controlling of diversity by a superior principle of unity, what Cage calls "finite temporal objects" (cf. Pierre Schaeffer's notion of the ideology represented in musical objects).

It would seem that today critics are able to discuss with greater clarity what I have called the wave-like nature of music. We can refer to Michel Serres, whose first studies of music found, in the work of Xenakis, for example, nothing more than signals which simulate background noise and are thus only (*necessarily*, according to Serres) a substitute for it.[9] But in the last chapter of *Esthétiques sur Carpaccio*, whose title alone ("Timbales et buccinateurs"), merits pondering, Serres advances a new theory:

> Music is, first and foremost, the ensemble of those messages which are minimally encoded (the *least* determined) for any given culture, the sum of the simplest and at the same time most highly polished signs. Hence its universal, transcultural appeal. Each *aire*, by means of artifacts or the human voice, produces and communicates this optimal series of waves. Hence its origin*ary* function in the crafts, the fine arts, the sciences (i.e., physics and the theory of elementary numbers).
>
> Music is precisely this minimum, this elemental and ultimate simplicity of signals which it puts into circulation. But what does the message carry? Nothing, strictly speaking. Not a single sign, or rather a block of arbitrary signs; we need only agree on the code. No meaning, or rather an arbitrary meaning; exactly whatever meaning one likes. Music is the communication of the commonplace, the ordinary. A pure signal which makes one's skin shiver, which overwhelms that ensemble of interpretative orifices which we call the body and makes one dance. An empty cell, it masks the relentless din of the world, blotting out the interminable noise of the real; it temporarily deafens us to that idiot background noise generated by the crazy, complex workings of our internal organs. These are almost pure signals, empty, neutral calls to the commonplace. There is a connection between this simplicity of encoding and that universal which the signal hollows out. Music is the encoded mode of communication of universals anterior to anything a message can transmit: a minimal degree of encoding, a generality of universals imparted. Better still, these universals are nothing other than themselves, they cannot define themselves in any way other than this maximum simplicity in the encoding of the signal.
>
> Background noise is the necessary condition for the transmission of any message: a basic multiplicity of sound which fills the channels and spaces of communication. The first, empty call is the necessary condition of every message to follow, as it is the least laden, so to speak, and hence the most capable of carrying an overload of meaning. These universals are the

[9] Cf. "Musique et bruit de fond," in *Hermès II, L'Interférence* (Paris: Ed. de Minuit, 1972), p. 191.

necessary conditions of knowledge, white empty forms to be filled. But they are not in some other world, or in a sealed black box called the "subject" or anything else, they are dispersed, here and there, locatable, flowing in space.[10]

If today's experimental music is indeed very concerned with voice and waves, then we might speculate, in tandem with McLuhan's theory that the technology of electronics is extending our brain and nervous system, that our minds are in the process of expanding our possibilities beyond known limits and that as a result we are beginning to perceive the globe and space that surrounds it WHOLLY, that is, from an ECO-LOGICAL or SOCIAL point of view and no longer from a purely IN-DIVIDUAL point of view. This is how Cage understands *satori*, or enlightenment in Zen Buddhism: one is enlightened when one begins to live in the world wholly, in a world which is not rational, techno-logical, or goal-oriented, but also irrational. As we have seen, our experimental music surrenders the "rational" organizing of sound to the higher laws of entropy and thereby takes an irrational, *non*goal-oriented stance toward the future. It discovers under the "unique" voice of the subject the "waves" of a crowd, a rhizomatic multidimensionality. This *"bruissement de la langue,"* as Barthes calls it, may enlighten us "not in terms of individual attainment, but in terms of social attainment."[11] Thus being responsible to the space that surrounds us, we have to behave ourselves MUSICALLY, we have to dwell within it POETICALLY, both *with* and *without* technology, saying both *yes* and *no* to it, in the spirit of equanimity which Heidegger describes as *Gelassenheit*.

Translated by Kathleen Woodward
University of Wisconsin-Milwaukee

[10]*Esthétiques sur Carpaccio* (Paris: Hermann, 1975), pp. 136–37.
[11]"John Cage Talks to Roger Smalley and David Sylvester," eds. Christopher Hobbs and David Sylvester, in Program Notes of *Music Now*, John Cage and David Tudor at the Royal Albert Hall, 22 May 1972.

Art and Technology:
The Panacea That Failed

JACK BURNHAM

Today's science has spawned a wealth of technical gadgetry, while on the other hand, modern visual artists have been notoriously unsuccessful in utilizing much of it in the making of socially acceptable art. Why should it be so? Some forms of technology seem to lend themselves to art which has gained museum status, yet even with the aid of millions of dollars in grants and private donations (plus the assistance of some of the biggest names in contemporary American art, e.g., Rauschenberg, Oldenburg, Warhol, Kaprow, Lichtenstein, Morris, and Smith), the results have fared from mediocre to disastrous when artists have tried to use what has euphemistically been referred to as the electronic technology of "postindustrial culture."

Precisely what succeeds in the context of art and what fails? Simple mechanical devices based on balanced catenary links such as Alexander Calder's mobiles or George Rickey's weighted blades seem to be the only kinetic sculpture fully accepted by the art world. In terms of luminous sculpture (which saw a dazzling revival in the 1960s), only Dan Flavin's unexotic fluorescent fixtures have gained permanent status in museum collections. Certain hand-manipulated objects such as the water boxes of Hans Haacke, the optical reliefs of Jesus Soto, and the Signals of the Greek Takis have some artistic validity. Curiously enough, the only machine-driven or electrically powered art that has maintained its status through the 1970s are the fantastic robots and constructions of the Swiss

200

Jean Tinguely, which are programmed in many instances to break down or malfunction. It must be remembered that during the 1920s Francis Picabia, Max Ernst, Man Ray, Marcel Duchamp, and Tristan Tzara joined in the systematic subversion of the machine as an artistic force. Moreover, one wonders if the Constructivist-Dadaist Congress in Weimar in 1922, was really an accident of accommodation as some of the participants later insisted, or if there was subconscious and interior motivation to the juxtaposition of Dada's brand of chaotic destruction with the mechanistic ideology of Constructivism. Why should the only successful art in the realm of twentieth-century technology deal with the absurdity and fallibility of the machine? And why should electrical and electronic visual art prove to be such a dismal failure?

At its ideological core, advanced technology has always maintained some of the chimerical effect that the perpetual motion machine had before the twentieth century; we are led to believe in its eternal stability, omnipotence, and its ability to perpetuate human enlightenment. We have been seduced into not doubting technology's efficacy because of its palpable short-term advantages. Yet why have the majority of artists spurned advanced technology, and why have others so bungled its use in producing new art forms? Is it possible that the schism between art and sophisticated technology is far deeper than we suspect, that, in fact, these differences may lie embedded in the neural programs of artists' and scientists' minds? Or are there teleological reasons for this schism, perhaps based on the theological foundations of the Judaic-Christian tradition? If so, let us review some of the recent evidence before surmising the reasons for it.

In Paris the dealer Denise Renée opened an exhibtion entitled "Le Mouvement" in 1953 with the help of K.G. Pontus Hultén and her partner Victor Vasarely. Included in "Le Mouvement" were Duchamp, Soto, Tinguely, Calder, Bury, and Agam. In March of 1961 the first "International Exhibition of Art and Motion" opened at the Stedelijk Museum in Amsterdam where it caused a *succès de scandale* for the organizers, in part because of the public response and the bitter tensions which prevailed between the Neo-Dadaists and the kinetic Constructivists. In April of that same year the Australian sculptor Len Lye mesmerized an audience at the Museum of Modern Art in New York City with an evening of "Revolving Harmonic" polished rods which created virtual forms at various speeds. Thus began a propensity for art in motion and light during the last decade which in 1967 *Time* magazine was to caption "The Kinetic Kraze." The rationale behind much of this esthetic was a simple one; namely, if so much of twentieth-century art was concerned with the depicted effects of light and movement, then why not produce art which literally relied on light and movement?

Until the early 1960s museums and galleries had tended to emphasize the historical aspects of light and movement. Technically this involved simple motor-driven devices, motorized light boxes, and various static light sources such as neon, incandescent and fluorescent fixtures. Following the Amsterdam Retrospective and an outstanding kinetic display at the 1964 Documenta III exhibition in Kassel, West Germany, the tendency moved towards an escalation of technical means, with a concurrent emphasis on collaborations between artists and research and engineering personnel. By the mid-1960s a division had developed between the earlier "machine art" and what could be defined as "systems and information technology." The latter includes artists' use of computer and online display systems, laser and plasma technology, light and audio-sensor controlled environments, all levels of video technology, color copy duplicating systems, programmed strobe and projected light environments using sophisticated consoles, and artificially controlled ecological sites. The definitive boundary line between the old and new technologies probably came with the New York Museum of Modern Art's 1968 exhibition "The Machine as Seen at the End of the Mechanical Age."

At this point it might prove beneficial to touch upon five major art and technology projects with which I have been tangentially or directly concerned. In some instances financial support or approximate budgets have been supplied. These are given to provide some yardstick with which to compare costs relative to standard museum exhibitions. If final evaluations for most of these projects appear overly negative, it should be remembered that these also express the general consensus of the art community and not just my opinion.

I. Experiments in Art and Technology

Dr. Billy Kluver, a Bell Telephone Laboratories' scientist specializing in laser research, had worked with top-level artists all over the world since the late 1950s when he had been an adviser for K.G. Hultén's kinetic exhibition in Amsterdam. In 1965, along with John Cage and Robert Rauschenberg, Kluver began to organize an art and technology extravaganza which became the ill-fated "Nine Evenings: Theater and Engineering," presented at the 69th Regimental Armory in New York City in October of 1966. Kluver, with the aid of some of the most prestigious names in American art, gained the support of some thirty patrons and sponsors amounting to over $100,000. The donated engineering aid was probably worth at least $150,000.

Each evening of "Nine Evenings" presented one or two uniquely designed "pieces," including large scale inflatable structures, radio-controlled dance vehicles, audio-magnified tennis games, infra-red projected "work tasks" performed in the dark, and complex musical pieces synthesized from a number of live external sources. On October 15 the theater critic Clive Barnes reported on the first performance of "Nine Evenings"; his view was more or less typical of the general audience response, particularly that of other artist spectators:

> If the Robert Rauschenberg work, "Open Score," had been a big and glorious fiasco—the kind of thing people write about in years to come rather than the next morning—it could have been a kind of little triumph. But in fact it was such a sad failure, such a limp disaster, more like an indiscretion than an offense. The level of the technology was such that the performance started 40 minutes late, a 15-minute intermission lasted 35 minutes and even a loud speaker announcement was so indistinct on the apparently unsound sound equipment that it became unintelligible. God bless American art, but God help American science.[1]

Barnes later pointed out that "Nine Evenings" was not so much an experiment in theater and engineering as it was an experiment in sociology, since it would take a particularly perverse audience to sit through and endure anything so feeble. Later defenders of "Nine Evenings," such as the critic Douglas Davis, alluded to the overall complexity and uniqueness of each performer's support system. "There was, to begin with," Davis has written, "the patchboard system. Each artist's performance was prewired; all of his equipment could be hooked up by inserting his particular patchboard. The system included amplifiers, relay decoders, tone-control units, transmitters and receivers; it also included a 'proportional control' network that made it possible to change the intensity and volume of both light and sound by moving a flashlight over sixteen photocells. . . ."[2] Kluver and his associates insisted that "Nine Evenings" had been a qualified success, based on the excellent rapport that developed between some artists and engineers working out problems on an intimate basis, and indeed, this has become the major rationale for claiming success for many subsequent art and technology mergers.

In January of 1967 Kluver and a group of associates published their first *E.A.T. News* bulletin as an outgrowth of "Nine Evenings." The public function of Experiments in Art and Technology Inc. was to act as a service organization, to make materials, technology, and engineering advice available to contemporary artists. Because of its governmental and

[1] Clive Barnes, "Dance of Something at the Armory," *New York Times*, 15 Oct. 1966, p. 88.
[2] Douglas Davis, *Art and the Future: A History/Prophecy of the Collaboration Between Science, Technology, and Art* (New York: Praeger, 1973), p. 69.

corporate ties, E.A.T. felt that it was in an ideal position to act as a liaison between artists and desired industries. Working from a Manhattan loft, E.A.T. held a number of seminars, lectures, and demonstrations for interested parties, and produced "Some More Beginnings" at the Brooklyn Museum in 1968. By 1970 Kluver and key members of E.A.T. had so proselytized on a nation-wide basis that according to their files they had upwards of 6,000 members, reportedly half artists and half engineers. No doubt, E.A.T.'s greatest success was its ability to extract relatively large sums of money from the National Endowment for the Arts, the New York Arts Council, large corporations, and various patrons of the arts. Technology seemed to be the key to loosening all sorts of purse strings. If business had been the business of the United States in the 1920s, surely in the 1960s the business of the United States was to acquiesce to the mystique of technology, as epitomized by the use of the "automated battlefield" and systems analysis during the Vietnam War.

The reputation of E.A.T. was irreparably hurt by its rupture with the Pepsi-Cola Company when it planned to produce an art and technology pavillion for *Expo '70* at Osaka, Japan. As Calvin Tomkins elaborates in his brilliant article for *The New Yorker,* "Onward and Upward with the Arts," the E.A.T. people, after many delays and financial fiascos in Osaka, presented Pepsi in April, 1970 with a maintenance contract for $405,000; the previously proposed sum had been $185,000.[3] Pepsi pulled out and E.A.T. gradually lost its image as a corporate mediator. Outside New York City, artist members of E.A.T. began to grumble that they were merely statistical fodder for E.A.T.'s grant proposals and that most of their serious requests to E.A.T. were simply ignored or bypassed with form letters. Once the word penetrated the art world that E.A.T. was an "elitist" organization, simply catering to the needs of its own staff and a few favored big-time artists in the New York area, its national demise was insured.

II. Cybernetic Serendipity

The first large-scale exhibition of "post-machine" art was held at the Institute of Contemporary Arts in London during the summer of 1968. Entitled "Cybernetic Serendipity," it was curated by Jasia Reichardt, an imaginative writer and vital force on the London art scene. Her catalogue-book contains a good layman's account of the historical development of digital computers, some relevant scientific projects, plus various experiments by artists that utilize feedback in machines. Other exhibits in "Cybernetic Serendipity" included computer printouts of musical

---

[3]Calvin Tomkins, "Onward and Upward with the Arts," *The New Yorker,* 3 Oct. 1970, pp. 83 ff.

analysis, computer-designed choreography, and computer generated texts and poems. But the I.C.A.'s exhibition was produced on a shoe-string budget: it did not use on-site computers or terminals and much of the available equipment was loaned. Moreover, when the exhibition was shipped to the Corcoran Gallery of Art in Washington D.C. the following year, a considerable portion of the contents was destroyed because of poor packing and handling. Several unpaid electrical engineers spent months salvaging parts of "Cybernetic Serendipity" for the opening, but Jasia Reichardt publicly disowned what was shown there.

III. Software

During the winter of 1969, Karl Katz, the director of the Jewish Museum in New York City, decided to mount a major exhibition based on computer technology and chose me to curate what was to become the first computerized art environment within a museum. "Software" did not open, however, until September of the following year. When I accepted, I hardly realized that the project would consume a year and a half of my life. Problems surfaced at every turn, ranging from dilemmas of conception and budgetary restrictions to malfunctioning of equipment and possibly even sabotage.

First, in planning the content of "Software," I was faced with an obvious quandary. At least two-thirds of extant "computer-art" consisted of computer programs designed to simulate existing art styles. Early on the use of the digital computer as a generative tool for creating art or music had been noted by Dr. John R. Pierce of Bell Labs. This was the case in the work of John Whitney, for example, who in the early 1960s began to program geometrical computer graphics using I.B.M. equipment. Similarly, Michael Noll had created a series of linear variations on known modernist masterpieces by using a line plotter. And there were many others: Kenneth Knowlton and Leon Harmon, Charles Csuri and Harold Cohen, to name only a few. But in spite of a wealth of official financial aid during the 1960s and early 1970s, most computer artists became profoundly disillusioned with the creative potential of tools. As Michael Noll admitted as early as 1970, "The computer has only been used to copy aesthetic effects easily obtained with the use of conventional media, although the computer does its work with phenomenal speed and eliminated considerable drudgery. The use of computers in the arts has yet to produce anything approaching entirely new aesthetic exper-ience."[4] And in fact, except for the magazine, *Leonardo,* edited by the ex-aeronautics engineer Frank Malina, the art world has been consistently unanimous in its refusal to recognize or in any way support computer-

[4]Davis, p. 111.

based art. With all this in mind, I decided with "Software" to forget about "art" as such and to concentrate on producing an exhibition that was educational, viewer interactive, and open to showing information processing in all its forms.

Sponsored by the American Motors Corporation through the agency of Ruder & Finn Fine Arts, a public relations firm, "Software's" initial budget was $60,000, not a princely sum, we were to learn, for an exhibition which expected to house four computers. The Jewish Museum expected substantial help from some of the smaller computer firms, companies specializing in software design, and various university departments that relied heavily on computer technology. I.B.M., we were told, was willing to pick up the tab for all of the exhibition's hardware and software. But the Museum and American Motors correctly perceived that "Software" would all too readily become a prime-time commercial for I.B.M. and thus the offer was rejected. However, two months before the opening of "Software"—with eight major computerized exhibits—we decided that an extra $15,000 was an absolute necessity to sustain the show through a two-month exhibition period. American Motors generously added this money to our budget of $60,000. And without the donated support of various corporations such as Digital Equipment Corporation, 3M Company, Interdata, Mohawk Data Systems, two members of the Smithsonian Institution design staff, and sundry individuals in the computer field, it is doubtful that "Software" could have been mounted for less than $25,000. Yet even after our major computer, the PDP-8, had been reprogrammed a second time, it took several D.E.C. engineers six weeks to make both "Labyrinth" (the interactive catalogue) and related exhibits operational. The computer's failure to function was a mystery to everyone and a source of embarrassment to D.E.C.

This was not the only operational difficulty. The day before "Software" opened, the exhibit which one encountered upon entering the show's space—a darkened pentagon of five film loops which showed artists working or explaining their conception of "Software"—was destroyed by two of the filmmakers themselves. Involved in a dispute over titling and finances with the producer of the films, they cut the five films to pieces; it took three weeks to resolve these problems and make copies from the master prints. And the night before "Software" opened, a janitor sweeping the floors of the Museum short-circuited the entire program of the PDP-8 by breaking some wires in a terminal stand with a push broom—or at least that was the official story released by the Jewish Museum.

The fact that "Software" opened without its film and minus the use of its central computer gave gleeful satisfaction to some members of the program of the PDP-8 by breaking some wires in a terminal stand with a

New York art press. The reasons for this animosity may stem from the ever-growing and disproportionate influence that technology exerts on our cultural values. As a result of training and personality, many art critics consider themselves "humanists" with strong feelings concerning the encroachments of technology on nature and cultural traditions. A few have successfully advocated what might be termed "Pop Technology," e.g., cybernetic light towers, video banks, and electronic sensoriums, but most critics instinctively realize that it would damage their art world credibility if they became serious advocates of hard technology as an esthetic life-style. With the rash of "Tek-Art" adventures during the 1960s, substantial numbers of artists and critics feared that electronics might soon overwhelm the prestige of the traditional art media as found in painting and sculpture. At the time, the spectre of an engineer-controlled art world seemed a bit too imminent for comfort. Hence, the reviews for "Software" were decidedly mixed, containing both strong praise and condemnation.

But on the whole, Talmudic scholars and rabbis situated on the top floor of the Jewish Museum were heard to mutter darkly as to the inappropriateness of exhibiting "Software" in a museum mainly devoted to Judaica and Jewish studies. The director of the museum, Karl Katz, lost his job a month after "Software" was disassembled. And the New York Trade Commission gave American Motors a special award in 1971 for sponsoring the most ambitious and interesting cultural failure of the year in New York City, an mixed blessing which American Motors, nevertheless accepted with gratitude.

IV. The Center for Advanced Visual Studies

One of the major attempts to wed art and technology in the United States during the last decade began formally in January 1968 with the opening of The Center for Advanced Visual Studies at Massachusetts Institute of Technology. Its founder was the head of the Visual Design Department at M.I.T., Professor Gyorgy Kepes, who in the early 1940s had headed the photography department at the Chicago Bauhaus under Laszlo Moholy-Nagy. Invited to M.I.T. in 1946 to organize the design program for student architects and engineers, Kepes created several important light murals during the 1950s and taught a seminar in 1957 on kinetic art, considerably before kineticism became fashionable in the United States. Possessing formidable connections within the scientific and academic world, he began plans for the Center in 1965. The Center for Advanced Visual Studies was to be the fulfillment of everything his mentor, Moholy-Nagy, had written about in his seminal *Vision in Motion* during the Dessau Bauhaus period. In 1967 M.I.T. renovated its old bookstore on

Massachusetts Avenue in Cambridge according to Kepes' plans. Essentially this consisted of five large, first floor studio areas, a large public work space in the basement, a small woodworking shop, plus a lavishly equipped photography darkroom.

In 1968 the German artist Otto Piene, the Greek sculptor Takis, Harold Tovish, Ted Kraynik, Wen-Ying Tsai, and I were invited to join the Center as its first fellows. Kepes' master plan for the Center was to produce a sophisticated environment where artists with a technological bent could do their own art and collaborate on large-scale group projects. In *Art and the Future*, Douglas Davis draws a fairly sympathetic portrait of Kepes' hopes and the early progress of the Center. Davis comments that the "Center's early years were lean ones financially, and that Kepes was kept from fulfilling his hopes in detail."[5] After a year at the Center my perception was at considerable variance with what Douglas Davis saw or believed.

Given the state of the American art world, Kepes initially had generous financial support, with M.I.T. and a half a dozen foundations backing him. But during the past few years support for the Center has dwindled as it has failed to produce writings, art works, or urban projects of any significance. Much of this is not the fault of the present director, Otto Piene, who has struggled to keep the Center alive. I would lay the blame in two directions: the rapid decline of technological art as one of the pet ideals of the avant garde, and the Center's lack of any concrete philosophy beyond the exploitation of available technologies. All too often artists expect their rather feeble art ideas to be rescued with the aid of exotic electronics.

Actually, except for those areas of scientific research that produced stunning photographs, such as holography, electron microscopy, and aspects of optical physics, Kepes had a strange aversion to direct involvement with sophisticated technology, particularly anything to do with the computer sciences. Due to the fact that the Center had been publicized, by virtue of its relation to M.I.T., as a technological nirvana for the artist, I found the situation mystifying. Slowly it began to dawn on me that the Center's underlying purpose was not primarily to do visual research or to make art, but to produce lavishly illustrated catalogues and anthologies that would impress foundations.

One should remember that in 1969 the Vietnam War and student-faculty protests were at their height. Speculation abounded that the Center was M.I.T.'s gesture towards the humanities, perhaps a means of focusing attention away from the presence of so many Navy and Air Force contracts. Certainly the Center never really had any concrete program,

5Davis, p. 115.

outside of fulfilling the director's vague dreams of creating urban spectaculars. During my first month and a half we met twice weekly to discuss Kepes' ambitions for erecting a colossal light tower in the middle of Boston Harbor. Somehow the conversations and exchange of ideas remained maddeningly vague. I began to ask specfic questions:

Did the Center have funds for such a project or any idea of costs? *No.*

Given that the Boston Harbor was directly in the flight patterns of Logan Airport, had the Center checked on the feasibility of the project with the local Civil Aeronautics Board, or with the Boston Harbor Authority? *No.*

Did they understand the problems of laying underwater electrical conduit or the costs? *No.*

What was the civic purpose of the light monument? *No one really knew.*

V. Art and Technology

Of all the art and technology projects instigated during the 1960s, Maurice Tuchman's five-year symbiosis at the Los Angeles County Museum (1967–71) was the most ambitious and perhaps the most revealing. In 1968 I visited the Los Angeles County Museum at the invitation of Tuchman, the Museum's Curator of Modern Art, in the capacity of consultant. From the start there was something grossly immodest about "Art & Technology" or "A & T" as the Museum called it. Tuchman managed to induce thirty-seven corporations in the Southern California area to contribute financial and technical support to resident artists. After three years of selection and various labyrinthine transactions which are documented in the "A & T" catalogue, the Museum came up with twenty-two artists who were paired to work with specific corporations. Out of these twenty-two artists, sixteen finally produced usable pieces or environments of the exhibition. Originally Tuchman proposed that the Museum contribute $70,000 towards supporting "A & T," while corporations, he felt, would contribute $140,000 in cash donations. By the Museum's own reckoning, its final budget was $140,000 for the expenses of "A & T," including three months of operating expenses. In terms of nonmonetary contributions by corporations, including materials, technical assistance, and the use of working facilities, I suspect the total outlay for "A & T" was between $500,000 and $1,000,000. If "Art & Technology" had been a critical success, or if its extravagance had not been so attacked by critics, quite likely the published budget would have been considerably higher.

By drawing up contracts for artists and supporting corporations,

Tuchman made certain that there would be no abrupt pull-outs, inadequate technical assistance, or failures to furnish length of exhibition maintenance for artists' projects. In retrospect, the technical support for Los Angeles' "A & T" exhibition was probably the most thorough and proficient ever supplied for an exhibition of its kind. And yet the length and legal binding character of "A & T's" contract was a facet of the project which critics attacked with vigor. Critics saw it as a covenant between two capitalist organizations (e.g., the museum and each of its corporate benefactors), in collusion with or against all the artists involved. Even Tuchman in the catalogue intimated that most of the artists in the show would not have participated by 1971, the year "A & T" finally opened, primarily because much of the art world believed by then that there was or is a nefarious connection between advanced technology and the architects of late capitalism. In the press "Art & Technology" was decimated, and not altogether for unsound reasons.

In a review of "Art and Technology" for *Artforum*, I tried to place the exhibition in an historical perspective that would make the responses of the art world more discernible:

> No doubt "humanist" art critics are going to pan A & T as another marriage of convenience with industry that fails to measure up to Henry Geldzahler's exalted view of the last 30 years. However, like Dr. Johnson's remarks on the virtues of singing dogs, defending A & T as the "best exhibition of its kind" is also questionable. In any case, due to the particular sociopolitical malaise that has gradually engulfed the United States, this show probably will be the last technological attempt for a while. If presented five years ago, A & T would have been difficult to refute as an important event, posing some hard questions about the future of art. Given the effects of a Republican recession, the role of large industry as an intransigent beneficiary of an even more intractable federal government, and the fatal environmental effects of most of our technologies, few people are going to be seduced by three months of industry-sponsored art—no matter how laudable the initial motivation. Certainly painting and sculpture do nothing to alleviate these conditions, but at least they are less exasperating since they avoid unpleasant juxtapositions.[6]

[6]Jack Burnham, "Corporate Art," *Artforum*, Oct. 1971, p. 67. This review appeared in *Artforum* along with a piece by Max Kozloff under the general heading: "The 'Art and Technology' Exhibition at the Los Angeles County Museum (Two Views)." What was interesting about this review was that both the Los Angeles County Museum and *Artforum* had asked me to write it—the former, I thought, because they trusted my objectivity, and certainly there was much of a critical nature in the 5,000 words that I wrote. But unknown to me, John Coplans, then Managing Editor of *Artforum*, sent out his most trusted critic. Max Kozloff's piece, "The Multimillion Dollar Art Boondoggle," *Artforum*, Oct. 1971, p. 72, was probably the most vicious, inflammatory, and irrational attack ever written on the art and technology phenomenon. It posed the Museum, Tuchman, and most of the artists connected with "A & T" as lackeys of a killer government, insane for new capitalist

One might look again at the large corporations supporting technological art and the artists receiving their sponsorship and conclude that both were guilty of some degree of naiveté, but hardly collusion for political purposes. While E.A.T. and other art groups held out the boon of "new discoveries" to corporations funding them, most companies were cynical and wise enough to realize that the research abilities of nearly all artists are nil. What companies could expect is a limited amount of good press for appearing "forward looking." To be sure, sociologists and several conceptual artists such as Victor Burgin and Hans Haacke have shown that pervasive philanthropy and museum-controlled "taste-making" do exert long term political control over the artistic tastes of the public. But given the costs and popular failure of technological art, it would appear an enormously inefficient means of swaying the masses, much less a means of promoting Technocracy as a successor to Capitalism.

In retrospect one could divide the artists participating into three categories: the techno-artists such as Robert Whitman, Rockne Krebs, Newton Harrison, and Boyd Mefferd who were esthetically allied with the light and kinetic movement; New York "name" artists such as Claes Oldenburg, Roy Lichtenstein, Richard Serra, Tony Smith, Andy Warhol, and Robert Rauschenberg who were only tangentially connected with art and technology; and finally the oddballs such as James Lee Byars, Ron Kitaj, and Oyvind Falstrom who provided the show's element of serendipity. The "name" artists tended to do enlarged or elaborate variations of their standard work or to cynically build into their projects hints about the utter futility of technology as a humanistic endeavor. Yet, as I stated in my review, by its nature art depends upon social compliance and cooperation; every successful artist places his or herself in the hands of the financial establishment: "Whether out of political conviction or paranoia, elements of the Art World tend to see latent fascist esthetics in any liaison with giant industries; it is permissible to have your fabrication done by a local sheet-metal shop, but not by Hewlett-Packard."[7]

The examples given so far—"Experiments in Art and Technology," "Cybernetic Serendipity," "Software," The Center for Advanced Visual Studies, and "Art and Technology"—are a representative cross-section of major art projects concerned with advanced "postindustrial" technology during the past ten years. Have they failed as art because of technical or esthetic incompetency, or because they represent some fundamental

conquests in South East Asia. Kozloff depicted half of the artists involved as "fledgling technocrats, acting out mad science fiction fantasies"; the more sophisticated artists he envisioned as cynical opportunists.

[7]Burnham, "Corporate Art," pp. 66−67.

dissimilarity as systems of human semiosis? Although it is clear that technical incompetency is partially to blame, I would suspect the latter is a more fundamental explanation. My experiences with semiology and iconography lead me to believe that the enormous vitality and will-to-change behind Western art is in a sense an illusion, just as technology harbors its own illusionary impulses. Only within the past ten years have we begun to accept the possibility that technological solutions are not universal panaceas. Gradually but surely, much of it in unspoken terms, we are beginning to accept evidences that scientific research and technological invention have their boundaries. Such a speculation would have been nearly unthinkable fifteen years ago when scientific grants were plentiful and the avant garde was the key to artistic success. Perhaps technology is only a matter of man-made or artificial negentropy which, because of its enormous productive capacity and ability to aggrandize perception into convenient and coherent packages of "information," we perceive as invincible, life-stabilizing, all-meaningful, and omnipotent.

Since the scientific revolution, art has become a protected cultural sanctuary; as empiricism has gradually dominated everyday cultural values and academic standards, art has been transformed into a sort of necessary way-station for the expression of anti-social sentiments. It liberates the human spirit by its inability or reluctance to become acutely self-analytical, while at the same time art remains implicitly critical of everything around itself. One might conjecture that art remains a knife-edge or balancing fulcrum for the human psyche. By that I mean it encompasses all aspects of the psyche equally; mythic fantasy, technological skill, esthetic idealism, manual craftsmanship, a variety of contents, but most importantly an internal semiotic consistency which prevents it from becoming absorbed by other disciplines, no matter how powerful or persuasive. If there is a teleological function to art, quite likely it is to lead us back to our psychological origins, to exhaust our material illusions by forcing us to understand the reality of mythic experience, for myths are merely the mental constructs we devise for our perception of the world, having particular properties isomorphic with the physical world. Yet increasingly we sense the fragility of art, the fact that modern rationalism tends to denude it of its most precious characteristic, its "believability."

In 1968, my book *Beyond Modern Sculpture* was published. What made the book controversial was the prediction that inert art objects would eventually exhaust themselves as a means of cultural expression (that is, lose their powers of contemplative evocation for human beings). I suggested that the art world was rapidly moving from "object" orientation towards a "systems" orientation in its perception of mundane reality. The book ended with a prophecy:

The stabilized dynamic system will become not only a symbol of life but literally life in the artist's hands and the dominant medium of further aesthetic ventures. . . . As the Cybernetic Art of this generation grows more intelligent and sensitive, the Greek obsession with "living" sculpture will take on an undreamed reality.

The physical beauty which separates the sculptor from the results of his endeavors may well disappear altogether.[8]

In a sense *Beyond Modern Sculpture* validated itself in terms of some subsequent art; where it erred gravely is in its tendency to anthropomorphize the goals of technology. As with Norbert Wiener's comparison of the ancient Jewish myth of the man-made Golem with cybernetic technology, I envisioned the resolution of art and technology in the creation of *life* itself. Yet, in a most ironic fashion, something other than that has taken place. Presently and for the near future the science of artificial intelligence has produced nothing approaching life-like cognition, but merely pale imitations of it. The cybernetic art of the 1960s and 1970s is considered today little more than a trivial fiasco. Nevertheless, avant garde art during the past ten years has, in part, rejected inert objects for the "living" presence of artists, and by that I am referring to Conceptual Art, Performance Art, and Video Art. In the case of such artists as Chris Burden, Joseph Beuys, Christian Boltanski, James Lee Byars, and Ben Vautier, art and life activities have become deliberately fused, so that the artist's output is, in the largest sense, *life-style*. During his last years, Marcel Duchamp often insisted that art, after all, was only the process of "making." Thus, in a literal way, art objects are merely materials, the semiotic residue of the artist's activities. What we are seeing when we view art is a fusion of cognition and gesture; as the historical semiotic of art evolves, this becomes increasingly apparent. Gradually the art object destabilizes, imploding upon itself. What is left is a series of partitioned fragments of the entire art-making process.

In the long run, technology may, like art, be a form of cognitive bootstrapping, an illusionary form of conquest over the forces of Nature. Both are vaguely deceptive in that they hold out the possibility of human transcendence, yet they only lead us back to a point where we can understand how we are dominated by our own perceptual illusions. In technology the sense of mastery, manipulation, and "otherness" is a more implicit assumption than it is in art. The ritual-making aspects of art do not sever man so effectively from his natural origins. Ultimately, perhaps, the very weakness of art as a cultural force—its conceptual confusion and lack of utilitarian value—gives it its strength.

[8]Jack Burnham, *Beyond Modern Sculpture: The Effects of Science and Technology on The Sculpture of This Century* (New York: George Braziller, Inc., 1968), p. 376.

Any attempt to explain why art and the electronic technologies are mutually exclusive can only be conjecture. Possibly, though, the reasons for this schism are metaphysical and not technical. At its foundations art may be a cognitive discipline or exercise, one that steers us towards the most primitive regions of the human brain. Physically, the brain is a jelly-like gray mass composed of billions of neurons sending and receiving billions of weak electrical signals per second. Providing that art is primarily a form of self-understanding (*re-cognition*), it would seem likely that the principles behind the historical evolution of art contain an exclusion principle. By that I mean a principle which does not allow the esthetic-cognitive functions of the brain to accept an electronic technology as an extension of inanimate objects. In a sense a certain rapport or similarity exists between the brain and electronic technology, although analogies between the two at this time are very gross. Traditionally the esthetic aura or charisma of art has existed within a Pygmalion-like paradox: art "lives" although it remains consecrated in dead, inanimate materials. To challenge that paradoxical state may very well jeopardize the mythic consistency of Western art.

When one speaks of the "mythic consistency of Western art" many alternate possibilities come to mind. What I mean by that is Western art's semiological consistency, that fabric of "believability" in contemporary thought which has possibly been best defined in Roland Barthes' illuminating essay, "Myth Today." Barthes suggests, and I feel correctly so, that virtually everything is subject to mythic interpretation, hence the limits of myth are essentially formal, not substantial.[9] Does such a broad generalization define myth out of existence? Or does it suggest that the efficacy of mythic thought is far more culturally pervasive than our intellectual conventions allow? Barthes, of course, has been a strong advocate of the second position. For him myth becomes in a sense "background," the naturalization and depoliticization of everyday speech. This suffices, as with Barthes' examples, to explain the subtleties of patriotic posters, dress codes, or bourgeois rhetoric, yet it allows us insufficient insight into the dynamic vicissitudes of equally if not more complex phenomena such as art history.

Here one might suspect that the level of historical discourse (that carried on in works of art by artists and not scholarly analysis) is essentially anagogic, having to do with the unresolved purpose of Judaic-Christian culture at the highest levels. In such a case, the linguistic conventions of signified, signifier, sign, and referent revert back to their theological forms of Father, Son, Holy Ghost, and last but not least, Man himself. The mythic consistency of the Judaic-Christian tradition is

[9]Roland Barthes, *Mythologies*, trans. Annette Lavers (New York: Hill and Wang, 1972).

premised on a somewhat multiple assumption: namely, man cognizes by virtue of perceiving dichotomies, he acts triadically through the agency of signs, but he only comes to know himself by dissolving thought and action in the recognition of unity. The theological term "anagogic" also refers to the transformation of drives from the unconscious into constructive ideation, which is just about as succinct a definition of Western art as one could hope for.

As such, Western art leads a double existence. It operates as an unveiled and exoteric activity, taught pervasively in schools (usually badly) and subject to the most commercial exploitation. Yet it contradicts Barthes' everyday mythic invisibility because art by its very paradoxical nature (its near perfect resistance to economic, psychological, or sociological interpretation), openly signifies an apparent mystery concerning the fusion of spirit and matter. So at the highest level, secrecy and a code of concealment are imperative for its cultural survival.

Dialectically art moves in Western culture towards the disclosure of the human psyche, which I would interpret as the life force unhindered by ego and self-consciousness. Even this is accomplished paradoxically in that art appears to be constantly moving away from clarity and resolution, and towards chaos and materialism. Technology's mythic consistency is no less subtle, because it springs from the accrued conviction of the intellect's invincibility. In a sense it resembles the other side of the human personality: lacking the psychic acceptance of the artist, it places its *raison d'être* in empiricism, which tends to lead it towards its worst enemies, paradox and meaninglessness. Nevertheless, while art and technology show signs of mutual exclusiveness, at the level of anagogic significance they may actually be completely tautological.

IV
Cybernetics, Constraints, and Self-Control

Changing Frames of Order:
Cybernetics and the Machina Mundi

ANTHONY WILDEN

I. The Problematic

> . . . cruel Works
> Of many Wheels I view, wheel without
> wheel, with cogs tyrannic
> Moving by compulsion each other, not
> as those in Eden, which
> Wheel within wheel, in freedom
> revolve in harmony and peace.
>
> William Blake, *Jerusalem* (1820)

Cybernetics, allied with systems theory and with contemporary approaches in ecology, is not simply a science or a methodology, but more significantly, a point of view. If it seems astonishing that the epistemological position we have called "cybernetic" for some thirty years is still unfamiliar to many people, we must remember that, for anyone raised in what we may term the Newtonian-Cartesian *epistemology* of science and the one-dimensional, atomistic *ideology* of our society, the non-mechanistic cybernetic perspective may at first sound contradictory or unscientific, or even teleological and anthropomorphic.[1]

[1]Some of the detailed argument underlying the perspective of this paper was worked out during research on the National Science Foundation project: "The Design and Management of Environmental Systems," directed by W. E. Cooper and H. E. Koenig in the Department of Electrical Engineering and Systems Science at Michigan State University. Their help is

219

Cybernetics is explicitly a science of models. But apart from any concerns about the complex relationship between our social ideology and the contemporary discourse of science, the fact that cybernetic models have had their source in the study of *engineered* systems continues to create problems about the applicability of cybernetic (and systems) theory to the non-engineered reality which surrounds us.

W. Ross Ashby quite rightly pointed out in 1956 that his closed-system cybernetics did not owe its basic postulates and theorems to any other discipline. Unfortunately, this *ab initio* construction has rarely been true of the *application* of cybernetic principles to real, non-designed systems. Ashby was careful to avoid such misapplications, but as I have argued elsewhere, he quite unintentionally exemplified this problematic in several instances.[2] More recently, Jean Piaget has made some somewhat dubious associations between cybernetics, cognitive development, social structure, and his own "genetic" or "constructivist" structuralism.[3]

We have also seen anti-Freudian psychotherapists attempting to use a systems-cybernetic perspective within what appears to be the neo-Freudian epistemological and ideological framework of the "autonomous ego."[4] Sometimes we find the traditional Cartesian dichotomy between the subject and the object hidden behind the expressions "system" and "environment." In economics, we may find in the literature examples of cybernetically inclined researchers apparently rewriting Adam Smith's *Wealth of Nations* (1776) without realizing it—without realizing that their conception of economic "self-regulation" is no less mysterious and providential than the moralistic model of the "hidden hand" underlying Smith's analysis of the surface structure of liberal capitalism.[5] Philoso-

gratefully acknowledged, as is financial assistance for continued research from the President's Research Fund at Simon Fraser University.

[2]See Anthony Wilden, *System and Structure: Essays in Communication and Exchange* (London and New York: Tavistock, revised edition, 1980), pp. 138–40, 371–74, and W. Ross Ashby, *An Introduction to Cybernetics* (New York: Wiley, 1969), pp. 271–72, 69. In the last analysis, however, it is not really of great significance what anyone *intends* or *means;* in the end it is what they *say* and *do* that counts. Any "calculus of intention" or motivation is in fact a psychological construct with no exit, for, from this perspective, all motivations are equal. Such a position is of little help in a scientific and critical understanding of human reality; its total circularity and self-closure make effective judgments impossible. It is in any case a usually unrecognized component of the ideology of "freedom" in our society (cf. Section 6)—and, as we know, the road to the reinforcement of the status quo is paved with good intentions. In dealing with writing, then, we do not deal with the author, but rather with the *text.* Otherwise we fall into what literary criticism has long called the "intentional fallacy."

[3]See Jean Piaget, *Structuralism,* trans. Chaninah Maschler (New York: Basic Books, 1970).

[4]See Paul Watzlawick, Janet Beavin and Don D. Jackson, *The Pragmatics of Human Communication* (New York: W. W. Norton, 1967).

[5]Cf. Garry Wills, *Nixon Agonistes: The Crisis of the Self-Made Man* (New York: Signet, 1969), pp. 216–30, 308–43, 532–34. Cf. also Anthony Wilden, "Piaget and the Structure of Law

phers, too, can fall into a similar trap, unconsciously using cybernetic terms to reformulate Hegel's rationalist and idealist dialectics of the "cunning of reason." Ethnologists and ecologists sometimes run into trouble with apparent isomorphies between distinct systems which turn out to be mere superficial similarities. This is especially the case when they attempt to cross the clearly defined boundary between natural and social ecosystems.[6] Influenced by systems analysis and systems engineering, some ecologists can be heard implicitly imputing a design to nature and to society, and consequently speaking in unrecognized teleological terms, rather than in consciously teleonomic ones. These problems call to mind the Providence that served as a guiding and controlling principle for the classical economists, as well as its historical corollary in biology, the beneficient Mother Nature who served with such regularity as an explanatory principle in the work of Charles Darwin.[7]

Moreover, before Magoroh Maruyama, amongst others, provided positive feedback (deviation amplification) with a proper place in cybernetic theory,[8] the almost exclusive concern of cyberneticians for mechanisms and processes implementing or displaying primary negative feedback (deviation suppression) made cybernetics more a science of control as such than a perspective that could also accommodate the various kinds of amplification and accumulation we find in the real world (e.g., the accumulation of productive capacity in our economic system over time).[9]

The crux of this problematic can be easily stated (which is not to imply that it can be easily resolved). Like other general theories, cybernetics, systems theory, and ecological explanation can usefully and legitimately be applied to complexes of system-environment relations differing widely in their *order* or kind of complexity and in the *scope* or extent of their organization. Any of these ecosystems will almost surely itself involve levels and kinds of complexity.[10]

The point is that if the theoretical approach applied to any ecosystem does not itself manifest a logical typing of orders of variety which closely approximates the logical typing of the hierarchies of variety in the eco-

---

and Order," in *Structure and Transformation*, ed. K. Riegel and G. C. Rosenwald (New York: Wiley Interscience, 1975), pp. 83–117.

[6]See Claude Lévi-Strauss, *The Elementary Structure of Kinship*, trans. R. Needham (Boston: Beacon Press, 1971), pp. 3–41. See also Anthony Wilden, *System*, pp. 233–73.

[7]See Stanley E. Hyman, *The Tangled Bank* (New York: Grosset and Dunlap, 1966).

[8]Magoroh Maruyama, "The Second Cybernetics: Deviation-Amplifying Causal Processes," in *Modern Systems Research for the Behavioral Scientist*, ed. W. Buckley (Chicago: Aldine, 1968), pp. 304–13.

[9]See Wilden, "Piaget."

[10]See M. C. Marney and N. M. Smith, "The Domain of Adaptive Systems," *General Systems Yearbook*, No. 9 (1964), pp. 107–31.

system under study, then the approach will be forced willy-nilly into an epistemological or ideological reductionism.[11]

This is a reductionism of complexity (and a consequent loss of information) quite different from the consciously methodological reductions necessary in scientific investigation. In essence it involves crossing boundaries between hierarchical orders of complexity in the ecosystem without realizing it. Lest the word reductionism conjure up misleading images of Procrustes' bed, it should be emphasized that this kind of reduction of the "data" to fit the approach is not quantitative, but qualitative; and that it does not simply involve the adjustments in a single dimension which occupied Procrustes, but rather an unrecognized switching *between* dimensions. The consequence is a "flattening out" of the various orders of complexity in the ecosystem, orders which in living and social systems are always arranged in hierarchies or heterarchies.

As an illustration of this one-dimensionalization, we may take Edmund Burke's notorious dictum: "The laws of commerce are the laws of Nature, and therefore the laws of God." Similarly, Paul Samuelson, in *Economics*, does much more than simply mix his metaphors:

> Takeovers, like bankruptcy, represent one of Nature's methods of elim-inating deadwood in the struggle for survival. A more efficiently responsive corporate society can result. But, without public surveillance and control, the opposite can also emerge. The Darwinian jungle is not guaranteed to produce a happy ending.[12]

My point, however, is not to fill this paper with a catalogue of examples. I would rather approach the relationship between cybernetics and society by means of a historical parable. This, I hope, will indicate more by the multiplicity of evocation than by the simplicity of criticism what I see as the problematic of cybernetics today.

That this should be a historical approach is essential to the argument, because it is not enough that systems-cybernetic models should be, as it were, "nondisciplinary" in Ashby's well-taken sense.

[11]For the term "logical typing," see Bateson's use and interpretation of this concept taken from Russell in *Steps to an Ecology of Mind* (Los Angeles: Chandler; New York: Ballantine, 1972), pp. 279–308; or the discussion between Alice and the White Knight about the song he is going to sing, in Carroll's *Through the Looking Glass* (Chapter 8). The discussion concerns the *name* the song is *called*, the *name* of the song, *what* the song is called, and what the song *is*. The essence of the the problem of logical typing is that whereas the relationship between items of the same logical type is perfectly straightforward, the relationship between logical types themselves is not, and is usually the ground of paradox. Cf. Wilden, *System*, pp. 12–18, 110–24, 185–88, 414, and Anthony Wilden and Tim Wilson, "The Double Bind: Logic, Magic, and Economics," in *The Double Bind: The Foundation of the Communicational Approach to the Family*, ed. C. E. Sluzki and D. C. Ransom (New York: Grune and Stratton, 1976).

[12]Paul A. Samuelson, *Economics*, 8th ed. (New York: McGraw-Hill, 1970), p. 505.

The systems-cybernetic perspective must be both interdisciplinary and transdisciplinary, and both semiotic and dialectical. On the one hand, in order to both uncover and bridge the gaps between the disciplines, it must be philosophical, historical, communicational, ecological, economic, and anthropological. On the other hand, in the hope of introducing new order into the currently disordered state of the discourse of science, the perspective must develop a semiotic vocabulary abstract enough—of a high enough logical type—to allow translation into the specifics of any given area of study. The vocabulary may well be borrowed from various disciplinary dialects, but it must in its use transcend the limitations of a jargon.

Lastly, it must also attempt to transcend its academic origins by having some manifest instrumentality in what Marx called "the language of real life": the day-to-day communication (*Verkehr*) of the socioeconomic system. It must in other words be ultimately translatable into the vocabulary of that complex of relations to which no single term does adequate justice: the production, reproduction, and exchange relationships of the classes, the races, and the sexes within-society-in-history-in-organic-and-inorganic-nature (socioeconomic ecosystems).

II. Models and Metaphors

> Concepts that are basically shorthand for process elude verbal definition.
> It is only that which has no history which can be defined.
>
> Nietzsche

Whether a society speaks science or myth or both, its single concern about the realm of ideas or ideology must surely be that its dominant epistemology and ideology perform a long-range adaptive function. This adaptive function must of course match the long-range survival values embodied in the material reality of human interaction which permitted the system of ideas to arise in the first place. Whatever other functions they serve, science and myth must ultimately be viewed as instrumental aspects of the systemic organization which they serve (but do not directly control). Roy A. Rappaport has shown conclusively in his now classic cybernetic-ecological study of the Tsembaga in New Guinea that this kind of matching of long-range survival functions in myth, ritual, and reality has indeed existed.[13]

Other ecologically-oriented anthropologists have provided evidence to suggest that if there is one feature which distinguishes capitalist and state capitalist societies from other kinds, it is that our kind of socio-

---

[13]Roy A. Rappaport, *Pigs for the Ancestors* (New Haven: Yale Univ. Press, 1968).

economic organization, with its accompanying ideology, is the exception rather than the rule.[14] On one level, what seems to be exceptional about us—although I cannot argue the case in detail here—is that our socio-economic organization appears to have passed from adaptivity to counteradaptivity within a single century or so (Section 8). At another level, what makes us peculiar is that the ideology which has become increasingly dominant in our society since the capitalist revolution has not only never matched its socioeconomic reality, but is now also in the process of trying to catch up with the present and the future by valorizing various aspects of the unrecoverable past.[15]

This valorization extends all the way from the current and profitable epidemic of nostalgia—"The great thing about the good old days was that we didn't feel nostalgic"—through the resurgence of forms of utopian and agrarian socialism and individualistic anarchism, along with their various religious equivalents, and eventually to demands for that economic condition most dreaded by the progressive establishment of the nineteenth century: the steady state economy.[16]

This said, we may be permitted to go back to re-examine a venerable and essentially cybernetic construction of social and natural reality without being nostalgic about it: the theological-alchemical model of the cosmos in the middle ages.

In a celebrated passage from *On Learned Ignorance* (1440), Nicholas of Cusa restated a medieval definition of God's locus in the cosmos that had been an intellectual commonplace since at least the time of Alan of Lisle (d. 1202) and Jean de Meun's contribution to the *Roman de la rose* (c. 1277):

> In consequence, there will be a *machina mundi* whose center, so to speak, is everywhere and whose circumference is nowhere, for God is its circumference and its center, and He is everywhere and nowhere.[17]

[14]See Marshall Sahlins, *Stone Age Economics* (Chicago: Aldine-Atherton, 1972); Maurice Godelier, "Anthropology and Biology: Towards a New Form of Co-operation," *International Social Science Journal*, 26 (1974), 611–35, and "Modes of Production, Kinship, and Demographic Structures," *ASA Studies* (London: Malaby Press, 1975); Andrew P. Vayda, ed., *Environment and Cultural Behaviour* (Garden City, N. Y.: Natural History Press), 1969; and Karl Polyani, Conrad M. Arensberg, and Harry W. Pearson, eds., *Trade and Market in the Early Empires* (Chicago: Gateway Press, 1957).

[15]See Wilden, "Ecology, Ideology and Political Economy" (Mimeograph, Simon Fraser Univ., 1975).

[16]Cf. Herman E. Daly, ed., *Toward a Steady-State Economy* (San Francisco: W. H. Freeman, 1973).

[17]This notion is restated in his self-consciously mystical *The Vision of God, Or The Icon* (1453). It is a modification of Alan of Lille's seventh "theological rule": "God is an intelligible sphere (*sphaera intelligibilis*), whose center is everywhere (*ubique*), and whose circumference is nowhere (*nusquam*)," derived from the pseudo-Hermetic *Book of the Twenty-Four Philosophers,* and known to Alexander of Hales, Vincent de Beauvais, Bonaventura, and Thomas Aquinas, amongst others. See Ernst R. Curtius, *European*

Cardinal Nicholas was a mystic, a mathematician, a philosopher, a theologian, and perhaps the first experimental biologist; he studied the respiration of plants. He, with his colleagues, probably understood better than many of us today in just what sense the scientific discourse of any age or culture is, in the last analysis, only a moderately consistent code of metaphors, amenable to different translations in different times and places.

We have no difficulty in giving Cusa's description of God's relationship to the structure (*machina*) of the cosmos the serious examination it deserves. It is not of course a statement about God, but rather a statement about the dominant structure of medieval socioeconomic organization. It invites several interrelated levels of analysis, all of which, it seems to me, both require and illuminate the systems-cybernetic perspective.

Like his forbears and contemporaries, Nicholas of Cusa understood his society as an ordered subsystem operating within an organically organized whole governed by a *hierarchy of constraints*.[18] The ultimate constraint on all communication in the system (production, reproduction, exchange, maintenance, interaction) is embodied in a mysterious principle called God. No subsystem in the whole can "go outside" or transcend this constraint of constraints without becoming effectively extinct.

Taking the metaphor seriously, *as a metaphor*, we note at once that "God" simply symbolizes the ultimate constraint on all past, present, and future behavior on this planet, the constraint we now call the principle of entropy. As A. S. Eddington implied in *The Nature of the Physical World*, published in 1928, entropy has always had a somewhat magical and mystical aura about it. He compared it to the type of thinking we associate with 'beauty' and 'melody.' Entropy is not simply the quantitative result of a qualitative punctuation of reality (as most so-called 'laws' of science before quarks are). It is rather an essentially qualitative conception, one which depends on the qualitative signification of a chosen relation between 'order' and 'disorder.' Like 'beauty,' entropy is a feature of

---

Literature and the Latin Middle Ages, trans. Willard P. Trask, Bollinger, 36 (New York: Pantheon, 1953), p. 353. It was an aphorism favored by the anti-Cartesian Pascal (1623–1662), split as he was between the "old faith" and the "new science." The concept has been analyzed in terms of the philosophy of science by Alexander Koyré, *From the Closed World to the Infinite Universe* (New York: Harper Torchbooks, 1958), pp. 5–27.

[18]Hierarchies of complexity, hierarchies of rules and metarules, hierarchies of constraints, and so forth are not to be confused *per se* with socioeconomic hierarchies involving people. In historical terms, socioeconomic hierarchies rise and fall, or undergo metamorphosis over time. They are consequently to be regarded as special cases of hierarchy: as *heterarchies* (i.e., as systemic networks in which the dominant locuses of constraint and control immanent in the system may change place and function—and their relative logical typing—in the overall structure through time). See Wilden, *System*, p. 248. See also note 21.

arrangement. But, unlike beauty, entropy can also be readily communicated about in the quantitative system of communication in physics, i.e., mathematics.

The parallel between the ideological conception of God in the middle ages and the epistemological conception of entropy in cybernetic and systems approaches is even more marked. For Nicholas, God is the locus where there is "the highest concordance." It is the locus of the coincidence of antitheses (*coincidential contrariorum, oppositorum*), the concordance of differences (*De concordantica catholica*, 1433). This might well be a statement about a system at maximum entropy (complete randomness), for at entropy all differences are equal. A system of random diversity cannot be distinguished completely from a determined single-state system (unity).

Even without this particular parallel, the concept of a hierarchically constrained universe is an elementary ecosystemic principle, and one that most readily distinguishes the rules governing the behavior of living and social systems from the linear or efficient causality which is assumed to operate in the isolated, mechanical, equilibrium systems of a Newtonian universe.[19]

III. Ideology and Organization

> Every sign . . . is a construct between socially organized persons in the process of their interaction. Therefore, the forms of signs are conditioned above all by the social organization of the participants involved and also by the immediate conditions of their interaction. When these forms change, so does sign . . . [Therefore] 1) ideology may not be divorced from the material reality of sign (i.e. by locating it in the "consciousness" . . .); 2) The sign may not be divorced from the concrete forms of social intercourse . . . 3) Communication and the forms of communication may not be divorced from the material basis.
>
> V. N. Voloshinov, *Marxism and the Philosophy of Language* (1929)

The ideological and epistemological construct medieval and early Renaissance society called "God" performed an essential and recognized socioeconomic function for the survival of the system. It performed a function which the constraint on growth we call the order/disorder

[19]Cf. Ashby, *An Introduction to Cybernetics*, p. 127; and Bateson, *Steps to an Ecology of Mind*, pp. 405–15.

relations of (planetary) entropy has not so far been permitted to do in our form of economic organization.[20] Like the "heavenly influences" in the pantheon of the alchemists, the astrologers, and the natural magicians, the "God" of the theologian and the mystic is simply a set of ideological *representations*[21] of the real and material conditions which govern the *mode of production* dominant in medieval times and the *reproduction* of medieval society itself.[22]

As in any social system of any kind, these ideological constructs *re-present* to consciousness (and to the unconscious), at the level of ideas and images, the *semiotic* organization of the medieval economic structure by the *information* flowing through the system at all levels. The various levels and types of information (semes or signs) which both (positively) control and (negatively) constrain the behavior of every individual subsystem in the whole are encoded in various sign-systems: in money, in capital, in commodities, in artefacts, in the patterning of the natural landscape by human activities, and in the structure of social relations.

Neither God nor the dominant ideology he serves are the actual principles of socioeconomic organization in the system (anymore than any ideological construct ever is in any known social system). They do, however, function as a set of *metastatements* about the actual organizing principles at work at the level of the *deep structure* of the system. The ideological metastatements are available to consciousness in verbal and nonverbal forms: in speech, in concepts, in images, in art, in writing, and

[20]See Wilden, *System,* and "Ecology."

[21]As Althusser usefully summarized it, "An ideology is a system (with its own logic and rigor) of representations (images, ideas, or concepts as the case may be) with a historical existence and a role in the heart of a given society. . . . As a system of representations, ideology is distinct from science in that the practical-social function is more important in it than its theoretical or knowledge function. . . . An ideology is profoundly *unconscious.* . . . It is above all as *structures* that these representations are imposed on the immense majority of people, without passing through their consciousness. They are cultural objects which are perceived/accepted/submitted to, and they act functionally on people by a process which escapes them" (translation modified), in *For Marx,* trans. Ben Brewster (New York: Vintage, 1970), pp. 231, 233.

[22]Most complex socioeconomic systems display (coexisting) dominant and subordinate modes of production. Their relationship (their relative logical typing) becomes more evident in times of crisis or change. A mode of production consists of the *means of production* (the natural ecosystem, the human population, a given technology) and the *social relations of production* (i.e., the relations stemming from the way the components of the means of production are organized). The dominant mode of production in the middle ages hinges on the lord-serf relationship in agriculture, itself dominated by the production and exchange of *use values* rather than by *exchange value* as such. Subordinate to it one finds wage labor based on exchange value in the crafts and in the mines, for example. Capital exists also, but it has yet to become a commodity. All the components of the capitalist revolution exist, but only where the commoditization of capital, land, and labor is nearly universal and fully dominant (c. 1800) can we legitimately describe the economy as capitalist.

in many other aspects of the patterned diversity of our world. Their relationship to the deep structure of the system can be described as that of a set of *messages* to a *code*.

A code may be defined as a set of rules governing the permissible construction of messages in the system. A code mediates the relationships between the goalseeking sender-receivers that employ it. Thus a code or set of codes is the basis of the creative principle that makes messages and relationships *possible* in the first place, at the same time as the constraints embodied in a code make an even greater variety of qualitatively different messages and relationships *impossible* in the system as it stands.[23]

Such messages will remain impossible—and indeed unimaginable for the sender-receivers mediated by a given code—unless some particular combination of processes and events sets off a reordering or restructuring of the code. In this eventuality, existing or novel variety in the environment of the sender-receivers may consequently pass from the status of *noise* (disorder unacceptable or invisible to the system) to that of *information* (acceptable or useable order), and thus become incorporated in the coding arrangements of a qualitatively distinct system. In adaptive cybernetic systems, internally or externally generated noise does not therefore lead necessarily to breakdown. If a system can restructure its codes at relevant levels so as to generate new order out of disorder,[24] it will undergo an evolution or a revolution, as the case may be. In a properly dialectical sense, the system will have passed through an *Aufhebung* by means of which the old order forms the basis of and at the same time gives way to the new; and the whole, as a whole, survives.

IV. Steady State

> The Creator made man of all things, as a sort of driver and pilot, to drive and steer the things on earth, and charged him with the care of animals and plants, like a governor or steward subordinate to the chief and great King.
>
> Philo Judaeus (30BCE–50CE)

> There was a god, either maker or governor or both, of all this whole engine of the world.
>
> Thomas More, *Heresyes* (1529)

---

[23]See Wilden and Wilson, "The Double Bind."
[24]See Heinz von Foerster, "On Self-Organising Systems and Their Environments," in *Self-Organising Systems*, ed. M. C. Yovits, G. T. Jacobi, and G. D. Goldstein (Washington: Spartan Books, 1959); and Bateson, *Steps*.

We used to be reminded by Robert Browning that when "God's in His heaven, all's right with the world." But this cliché would hardly have been understandable at all for many people in the medieval context. Whatever we may now find objectionable in the feudal arrangements of the medieval *communitas*, perhaps the most significant aspect of its organic relationship to its interventionist deity was that, if God was anywhere, he was most certainly not simply in heaven, looking down on the world—as he was to be for post-seventeenth-century society. On the contrary, as the quotation from Nicholas of Cusa makes clear, even at the moment when the medieval world order was about to be transcended by the "commercial revolution" associated with the "age of discovery," God was still perceived by the theological establishment as *both* in heaven *and* in the world. There was no barrier or gulf separating humanity from God; rather there was an intimately linked "great chain of being" ascending through all the species from the lowest matter to the highest Form, to the absolute first and last represented by God as the *primum mobile*.[25]

In the words of the mathematician, philosopher, and economist Nicole Oresme (c. 1325–1382), in his commentary on Aristotle's *De Caelo*:

. . . God does not need the heavens or any other place for Himself because He is everywhere—both outside and inside the heavens. . . .
The heavens are moved by immaterial and incorporeal or spiritual things called intelligences or separate substances. . . . Spiritual things are indivisible and neither occupy nor fill any place. . . . Each intelligence is whole and wholly in every part—however small—of the heaven that it moves, just as the human soul is whole and wholly in every part of the human body—except that the soul is in the body by information (*informacion*), and an intelligence is in its heaven by mutual fitting together (*apropriacion*).[26]

What we immediately see in these images of a perfectly ordered world organism is another statement about the dominant pattern of production and social relations in medieval times. All relationships are conceived to be truly interrelational; the dominating pattern is that of a hierarchical unity of differences and a reciprocity of oppositions. Sympathy matches antipathy; difference is mediated by similitude; patterns are repeated through analogy and emulation; and the whole is bound together in the reciprocal coming and fitting together of mutual convenience. *Similitudo, analogia, aemulatio, convenientia*: the dominant theme is that every apparently separate entity shares its being—its "vibrations"—in some way with all the others; that they all know their place (in every sense); and that the cosmos is ordered as it is because all beings literally *co-operate* together in the whole.[27]

[25]Arthur O. Lovejoy, *The Great Chain of Being* (1936; rpt. New York: Harper Torchbooks, 1960).
[26]Nicole Oresme, *Du ciel et du monde*, fols. 69d–70a.
[27]See Michel Foucault, *The Order of Things: An Archeology of the Human Sciences* (New York:

The actual socioeconomic reality is less pleasantly harmonious, of course. It is true that in medieval society cooperation predominates over competition in production, just as the exchange of use values dominates the production and exchange of exchange values, but this cooperation is hierarchically enforced by the medieval class structure. Nevertheless, what does emerge from comparing the ideology with the reality is that medieval society approximated a steady state economy. The feudal mode of production was as exploitative as any other class-centered system. But it was not predominantly based on the maximization of gain through the institutionalization of competitive *uncertainty*, with the resulting necessity of continuous linear expansion, as is our own. On the contrary, except for the still subordinate activities of the merchant class, the feudal economy was based on the maximization of *certainty* through cyclic repetition. As represented as late as 1416 in the iconic images of the *Très riches heures du duc de Berry*, for example, the medieval economy was in essence a homeostatic system whose fluctuations were keyed more directly to the succession of the seasons than to its own generation of various types of disorder.

The rising merchant class was already shaking this system at its roots in the fifteenth and sixteenth centuries. By the seventeenth century, the still existing feudal mode of production was clearly occupying a subordinate status in the system. Its ideological representatives no longer form an integral part of the religious, political, and scientific establishments, except insofar as they play out their role as "conservatives," inveighing against technological innovation (Montaigne, John Donne), or resisting the expansion of credit (the repeated condemnations of usury).

In science, a century after the birth of Galileo, there is still a sufficient confusion about the pace and the actual characteristics of ongoing socioeconomic changes for Giambattista della Porta's *Magia naturalis*, first published in 1558, to be translated into English (1658). The steady state images of constraint remain—"The Sun is the Governor of Time and the Rule of Life," says Porta—but they have been overtaken by events. Western Europe is on its way towards breaking down or bypassing the traditional constraints on production; it is about to invent the uniquely modern ideology of "progress"—progress unlimited in time or space—to justify the economic necessities of expansion at the basis of the new order.[28]

Pantheon, 1970), pp. 17–44; Giambattista della Porta, *Natural Magick* (1658; rpt. New York: Basic Books, 1957). On the logical typing of cooperation and competition in social systems, see Wilden and Wilson, "The Double Bind."

[28]See Sydney Pollard, *The Idea of Progress* (1968; rpt. Harmondsworth: Penguin Books, 1971); Wilden, "Bateson's Double Bind," *Psychology Today*, Nov. 1973, 138–40; and Wilden, "Piaget."

V. The Locus of Control

> Systems [of explanation] in many re-
> spects resemble machines. A machine is a
> little system created to perform, as well
> as to connect together, in reality, those
> different movements and effects which
> the artist has occasion for. A system is an
> imaginary machine invented to connect
> together in the fancy those different
> movements and effects which are in
> reality performed.
>
> Adam Smith,
> *Essays on Philosophical Subjects* (1795)

Let us return once more to Nicholas of Cusa's image of a God who is both the center and the boundary of the system he has created. Precisely because he is mystically inclined, Cusa is looking far more deeply into the biosocial ecosystem than those amongst his predecessors and contemporaries who were content to conceive of God, government, and the emerging nation state on the model of the patriarchal family, as represented by the Church hierarchy and the Western monarchies, for example.[29]

If God is everywhere in the system, then God, the ultimate constraint, is not a controlling agency external to the system, as is a household thermostat, for example. In other words, the primary locus of constraint and control in this medieval system is exactly where it actually is in all non-engineered living systems: it is in the structural relations of the system itself. Like Cusa's God, structural relations are everywhere, but since (like memory) they cannot be located or localized, we may also say that they are nowhere at the same time. As in natural ecosystems and in most social ecosystems, constraint and control lie in the hierarchical and heterarchical networks of the system itself, *both* at the level of the individual subsystem *and* at the level of the whole.

This informational conception of the *immanence* of the locus of constraint and control in the relations between the "partials" of an ecosystem is one we in the West have had to rediscover in this century, once the Newtonian energy-entity equilibrium models of social and biological reality were found wanting. Not surprisingly, we find practically no trace of any similarly atomistic epistemology in Chinese science and ideology. In 1956, Joseph Needham noted the similarity between the geometric metaphor of the circle or sphere with its center everywhere and its circumference nowhere and the Chinese aphorism "Wu chi erh thai chi."

[29]See Donald A. Schon, *Invention and the Evolution of Ideas* (London: Social Science Paperbacks, 1967) for an analysis of the metaphorical remnants of this position.

We do not know precisely what this proposition meant, but it may be approximately translated as: "That which has no pole. And yet itself the supreme pole!" Needham compares the saying to the "Ungrund doch Urgrund" of the mystic Jakob Böhme (1575–1624). But Needham points out, correctly I think, that the Chinese proposition should be taken as a metaphor of the reality of Chinese society and its basic epistemological premises, rather than as a mere poetical, mystical, or theistic statement. The Chinese organization of socioeconomic reality has been for some thousands of years quite overtly systemic, ecological, and cybernetic.[30] The dominant epistemology in China never gave up the principle that the "order of things" was embodied and imbedded in the structure of natural and social reality. Neither nature nor society were considered to be ultimately governed by a potentially external principle equivalent to a divine rule or lawgiver. As in Western alchemy and natural magic, what ultimately ruled the cosmos was *Li:* Form or Pattern (information). Consequently, the Chinese never produced the essentially juridical notions of "natural law" and the "laws of nature," implying as they do a Creator or Lawgiver, and, as a result, an anthropomorphic *designer.*[31]

The Western association of creator, law, design, and external control (whether the controller is assumed to be God, the ruling class, or the government) has in fact been the greatest of ideological and epistemological obstacles to the proper understanding of systemic behavior, whether in nature or in society. This is not the place to detail the socioeconomic reasons for this difference between East and West. Nor can I analyze here the paradoxical but explicable fact that an anti-ecosystemic epistemology, ideology, and pattern of economic behavior have been *necessary* for the short-range survival of capitalism for the past three centuries.[32] But we have every reason to suspect that the only explanation of why we are now discovering these ecosystemic conceptions is that our economic system is working its way towards a crisis of such grave proportions that it both generates and needs them.

---

[30]See Joseph Needham, *Science and Civilization in China,* II (Cambridge: Cambridge Univ. Press, 1956), p. 11. If Needham is correct in his interpretation of the "south-pointing carriage," the Chinese also invented the first closed-loop negative feedback device, antedating the level regulator for water clocks invented by Ktesibios in the third century BCE. See *Science and Civilisation in China,* IV, pp. 286–303.

[31]See Needham, II, pp. 288–91, 344–45, 460–65.

[32]Cf. Wilden, *System;* "Piaget"; and Wilden and Wilson, "The Double Bind."

VI. Determinism and Free Will

> [The factory system] involves the idea of a vast automaton, composed of various mechanical and intellectual organs, acting in an uninterrupted concert for the production of a common object, all of them being subordinated to a self-regulated moving force. . . . Three distinct powers concur to their vitality [manufactures]—labour, science, capital; the first destined to move, the second to direct, the third to sustain. When the whole are in harmony, they form a body qualified to discharge its manifold functions by an intrinsic self-governing agency, like those of organic life.
>
> Andrew Ure,
> *The Philosophy of Manufactures* (1835)[33]

Taken as such, the underlying organic model used by Ure in his description of the early modern factory seems if anything more holistic and less mechanistic than Nicholas of Cusa's *machina mundi*. In reality, of course, the opposite is the case. Whereas *machina* could once stand as an equivalent of "structure" or "frame," and "engine" still retain its semantic connections with the Latin *genius* (spirit), the word "organ" now stands for a machine or for the alienated human beings attending it. The natural and social ecosystems, as well as their inhabitants, could now indeed be profitably represented in science and ideology as Newtonian machines, driven by the linear and efficient causality of a one-dimensional universe.

It comes as no surprise that the nineteenth century invented the word that most aptly labeled the machine perspective, the word "determinism" (as distinct from fatalism). After the Great Depression of the 1840s, "determinism" became almost the equivalent of a political slogan in the ideological and epistemological conflicts stemming from the competitive struggle for economic and political power in nineteenth-century society. Insofar as it represented science (physics, biology,

---

[33]Andrew Ure, *The Philosophy of Manufactures* (London: C. Knight, 1835). The supreme advantage and indeed the principle of the factory system, explains Ure, is that it allows children, trained to superintend a single "self-regulating mechanism," to replace craftsmen. This "union of capital and science" will do away with the problem that "by the infirmity of human nature . . . the more skillful the workman, the more self-willed and intractable he is apt to become, and, of course, the less fit a component of a mechanical system, in which, by occasional irregularities, he may do great damage to the whole" (p. 20). In 1831, Ure patented a "self-acting heat governor," a thermostatic device based on the thermocouple.

evolutionary theory), determinism occupied one pole of the opposition
between science and religion. Insofar as it represented technological
"progress" achieved by understanding the "laws of nature," determinism lay on the side of the manufacturer against the landowner. And
insofar as determinism, along with mechanism, was the watchword of
the "progressives" in nineteenth-century science, it did battle against the
mystical, animistic, ethical, and religious attitudes summed up in the
expressions "teleology," "vitalism," and "free will."

From the systems-cybernetic perspective, however, the debate
between the determinists and the vitalists in biology, and between
determinism and free will in ethics, provides us with little by way of an
explanation of behavior in living and social systems. In the socioeconomic
and historical context which makes individuals possible—the context
which both provides them with humanity and socializes them to be what
they are—the activities of the individual person are *neither* determined in
the classic sense, *nor* are they the products of free will. The supposed
polarity between these terms or conditions is in fact an ideological
illusion, and the debate itself can be seen as representing one of the
basic—and paradoxical—relations inherent in social democracy.

If we are led to feel that we must choose to say that in society we are
either free *or* not free, then it makes no difference which pole of the
paradox we choose. Alone, each alternative is contradicted by experience.
Similarly with the either/or of determinism. As in all such questions, the
very act of choosing one alternative will require us to turn back to choose
the other, and so on *ad infinitum*. The question of free will versus
determinism is a double bind, a paradoxical injunction.[34]

The real question is not whether we are free or not free, but *what* we
are free to do, and the one great freedom ideologically guaranteed us by
the commoditization of labor under capitalism is that we should be free to
sell our labor potential (our creativity) at the best price, and with "equal
opportunity" (to compete).

The middle ages had similarly to deal with the relationship between
freedom and determinism, except that this was not the Newtonian
determinism of *efficient* causes, but rather the Aristotelian determinism of
final causes: the teleological determinism represented by God in his
omniscience. In the medieval context, however, this relationship does
not have the characteristics of a double bind. On the contrary, the
theological-alchemical model of the cosmos successfully neutralized any
potentially paradoxical relation between human will and God's will by
means of the doctrines of "faith" and "grace," which allowed the

---

[34]See Bateson, *Steps*, pp. 201–27, and 309–37; Wilden, "Bateson's Double Bind";
Watzlawick, Beavin and Jackson, *The Pragmatics of Human Communication;* and Wilden and
Wilson, "The Double Bind."

Christian to choose *both* sides of the question at the same time. Symbolized in the Trinity, with the "only begotten son" as the man who was God, the very idea that there could be two sides on such a question was itself absurd (e.g., the "*Credo quia absurdum*," attributed to Tertullian).

If we now translate the problematic back into the cybernetic interpretation of the medieval cosmos with which we began, however, we can see why a both-and choice, *not dependent on faith*, was indeed possible (as also in China). There is involved here only an apparent opposition between two poles, rather than a real one. The relevant terms are not in fact "free will" and "determinism," but "goalseeking" and "constraint," between which there is no necessary opposition or contradiction. Once translated in this way, the whole question disappears, to be replaced by a definition of an ecosystem in cybernetic terms. For an ecosystem is simply a "phase space" of hierarchically ordered constraints within which individual goalseeking systems are free to live and move and have their being. Individual goalseekers may follow any number of trajectories within this ordered space as long as no trajectory breaks the boundaries defined by a given level or set of constraints. (These constraints include those embodied in each subsystem's relationships with other goalseeking subsystems.) To go outside the boundary is to invite death or extinction—with this exception, nevertheless, that if in the process of breaking through any particular boundary, a restructuring or transcendence of its associated constraints takes place, then some aspect of the system will have undergone the morphogenesis of evolution or revolution, and it will be possible for a new system to emerge from the old.[35]

VII. The Splitting of the Ecosystem

I am come in very truth leading to you Nature with all her children to bind her to your service and make her your slave.

Francis Bacon,
The Masculine Birth of Time, Or the Great Instauration of the Dominion of Man over the Universe (1603)

The commands through which we exercise our control over our environment are a kind of information we impart to it. . . . In control and communication we are always fighting nature's tendency to degrade the organized and to destroy the

[35]See Wilden, *System*, pp. 353–77.

> meaningful: the tendency . . . for en-
> tropy to increase.
>
> Norbert Wiener,
> *The Human Use of Human Beings* (1954)[36]

This restructuring of constraints was precisely what was accomplished in the three or four centuries of change (c. 1500–1800) which eventually produced that particular morphogenesis in the deep structure of Western society we now label the capitalist revolution.

The essence of this change can be captured in the metaphor of "infinite progress" (infinite growth, the infinite production and accumulation of exchange values). This restructuring required that the emerging system transcend the constraints on growth represented metaphorically in the medieval steady state economy by the figure of God. And we do indeed find the sign of this emergence quite starkly represented in the idea that the God who had once been immanent in the person and in the social universe (as in nature) had now decided to withdraw his presence from the world. The God who had always been, in one commonplace image, the stage manager of the *theatrum mundi*, now became a mere spectator.

This novel status, that of the "hidden God,"[37] is prefigured by Nicholas of Cusa in the way he employs an image from Isaiah to refer to God, but without any hint of alarm, as a *deus absconditus*. In contrast, by the time of Pascal and his fellow Jansenists living out the rapid scientific, technological, and social changes of the seventeenth century, the "Vere tu es deus absconditus"[38] had taken on a new and much more frightening meaning. The idea that God had abandoned all relationship to humanity and left it to its own devices—the implication that the coherence of the old

---

[36]See Benjamin Farrington, *The Philosophy of Francis Bacon* (Chicago: Phoenix, 1966), and William Leiss, *The Domination of Nature* (New York: Braziller, 1972), on Bacon's role as an ideological champion of the use of the new science of the seventeenth century in the (equally new) domination of nature. On the relationship between sexist images of the alienation of nature, the mind/body split, and other exploitative or oppressive dichotomies expressed in Western ideology, see Wilden, *System*, pp. 131, 217–25. C. S. Lewis once remarked that what we call "Man's power over Nature" usually turns out to be "a power exercised by some men over other men with Nature as its instrument" (as quoted in Leiss, p. 195). Note also in Wiener's remarks the one-dimensionalization of the logical typing of complexity in nature. In not respecting the boundary between organic and inorganic systems (DNA), Wiener fails to recognize that organic systems are not only the source and the ground of all meaningful organization, but that they are also responsible for its maintenance and reproduction. The entropy they create is continuously neutralized by the input of energy into the closed system of the biosphere by the sun. The cosmic or solar entropy Wiener refers to is irrelevant in human concerns, when compared with the increase in planetary entropy being produced by the global economic system (not by "nature").

[37]See Lucien Goldman, *Le Dieu caché* (Paris: Gallimard, 1955).

[38]"Truly Thou art a hidden God" (*Pensées*, Nos. 366, 591–99; Isaiah, 45; 15).

ecosystem had been split asunder—had become part of a growing tragic vision about the future of humanity.

Today, living as we do in an economic system which has discovered how to exploit all of its environments—geographic, human, natural, and temporal—and which continues to do so without limit, we necessarily remain much closer to Pascal's tragic vision of the splitting of the ecosystem and to Bacon's triumphal declaration of the war on nature than to the Chinese and medieval conception of the unity between and within society and the natural world.[39]

Pascal lived in a century which, unlike the nineteenth century, had little idea of where it was going. It took the political and economic crises of the twentieth century to allow intellectuals to rediscover the Pascalian feeling of "abandonment," the expression which became the watchword of existentialists like Heidegger and Sartre in the thirties and fifties. The deepening crises of the sixties and the seventies have only reinforced this feeling that somehow our society has lost its way. Looking back, we see that Pascal's "spectator God" was as good as dead. Looking forward, we may well suspect that the modern substitute for "God in his heaven"—I mean the utopia promised to all by progress through quantitative economic growth—is equally moribund.

If it is, we have no idea what to replace it with. We are not faced here with some sort of "eternal return" in history, a cyclic process which would allow us to assume that we face the same old problems and can therefore resolve them by applying the same old solutions. On the contrary, whatever problems the young capitalist revolution created for itself in the seventeenth century, it still had adequate flexibility—and adequate ecotime and ecospace—to go beyond them. Now that we live in a system which has apparently discovered the ultimate in exploitation, the exploitation of its own future and its own generations to come, we may well wonder whether the kind of systemic restructuring it is apparently heading for is even *imaginable* from our position within its ongoing processes, much less "engineerable."

VIII. Functional Controls and Structural Constraints

> As in a sudden flood, medieval constitutions and limitations upon industry disappeared, and statesmen marvelled at the grandiose phenomenon which they could neither grasp nor follow. The machine obediently served the spirit of of man. Yet as machinery dwarfed human

---

[39]See Leiss, *The Domination of Nature;* and Wilden, "The Domination of Nature," *Psychology Today,* Oct. 1972, 28–32.

strength, capital triumphed over labour
and created a new form of serfdom. . . .
Mechanization and the incredibly elaborate division of labour diminish the
strength and intelligence which is required among the masses, and competition depresses their wages to the minimum of bare subsistence. In times of
those crises of glutted markets, which
occur at periods of diminishing length,
wages fall below this subsistence minimum. Often work ceases altogether for
some time . . . and a mass of miserable
humanity is exposed to hunger and all
the tortures of want.

Fritz Harkot (1844)[40]

This question of future change returns us to the problematic with
which we began: the applicability of various approaches calling themselves "cybernetic" to the cybernetics of real life in living and social
systems. Manifest as it is in the work of many reformers and futurologists,
amongst others, we seem still to be saddled with one version or another of
the traditional "technocratic" perspective on socioeconomic change. This
viewpoint carries with it explicit or implicit assumptions about the
"designed" origins of social systems (respectably ensconced, for example, in the eighteenth-century version of the "social contract," derived
from Locke), as well as correlative projections about the effectiveness of
consciously engineered solutions to socioeconomic crises (as in the now-
forgotten flood of utopian, anarchist, and agrarian socialist writings
responding to the economic crises of the 1820s and 1840s, for instance).
As I have tried to point out, these assumptions and projections, in the
modern period, are intimately connected with the metamorphosis of the
metaphor of the Divine Creator and Lawgiver into that of the Divine
Artificer, and thence—by the mediation of the man-made machine
system—into that of the "Humanistic Engineer." (The other side of this
coin, and an enduring present conflict, might be called the metamorphosis of the Hidden God into the Laissez-Faire Economist.)

Whatever the apparent accomplishments of the recent past (and our
present economic system is a mere three centuries old), it seems unlikely
that the *long-range* adaptivity of the global economy—its long-range
adaptive stability in relation to all of its various environments—can be
maintained very much longer by the surface-structure adaptations and

[40]Quoted by E. J. Hobsbawn, *Industry and Empire* (Harmondsworth: Penguin, 1968), pp.
65–66. Harkort was a liberal German businessman and engineer; like Ure he represented
the rise of the modern technocrat. In most of its respects, this quotation could be directly
applied to the status of the Third World under global capitalism today.

"error-corrections" we are familiar with. The most obvious example of this type of adaptation—immediate "corrections" for immediate problems, with little concern for the whole structure of the system in its environments, and even less for the historical processes of which it is the result—is the relatively recent practice of the conscious "tuning" of the economy by governments and their agents. As even Keynes himself can be seen to have recognized, however, this surface-structure "tuning" may in fact amount to no more than a repeated application of unproven patent medicines to what are in actuality surface *symptoms* of structural contradictions and paradoxes developing in the system as a whole. The remedies are temporary and "physiological," rather than "morphological."

Moreover, other economists are now beginning to suspect that the repeated application of short-range remedies to long-range problems does not simply result in a failure to resolve many significant problems. The overall result is more invidious. Misguided attempts at solutions may actually aggravate the overall situation of impending crisis: short-range adaptations may in fact multiply over time—and as a result of their own effects—into states of long-range maladaptation or counter-adaptivity. Significantly enough, also, even the more grandiose schemes concerned to avoid what are perceived to be approaching socioeconomic and/or ecological disasters usually share the now traditional "technocratic" attitude. Instead of examining the question of the *restructuring* of the system, these approaches tend to concern themselves only with the surface-structure question of the *redistribution* of the present inputs and outputs of the system, leaving its fundamental organization, and of course its ideological and economic values, essentially untouched, as if they were *beyond question*.[41]

[41]See Arthur Shaw and Donald Sposato, "Implications of an Alternative World Exchange System," *Transactions of The New York Academy of Sciences*, 35, No. 7 (1973), 557–72. Apart from the fact that their cybernetic model is concerned primarily with economic *exchange*, rather than also with the fundamental processes that constrain all exchanges—the production of commodities and consumers, and the reproduction of the commodity system itself—Shaw and Sposato use a telling analogy from Wiener's writings: his comparison of the economic process to the game of Monopoly. Monopoly is a zero-sum game of real-estate speculation, not a representation of the *non-zero-sum* activities of real monopolies in the economic environment of real production and consumption. (In the short-run, at least, corporations are well aware of the fact that in any significant zero-sum competition with their multiple environments—a situation in which the *global* socioeconomic system may well find itself— the "winning" system is necessarily doomed to the equivalent of extinction, after the event.) However familiar to a particular class of people it may be—the class of home-buyers and mortgage holders—real estate speculation is a minor and subordinate component of our mode of production. The comparison, moreover, is not an innocent one, any more than Monopoly is an innocent game. The "dog-eat-dog" values of Monopoly are undoubtedly useful in the reproduction of the dominant ideology of "free and equal competition"—along

The well-known Meadows' study, *The Limits to Growth*,[42] for example, employs an overtly systemic and cybernetic approach to the problem of the various environmental limits on economic growth on this planet. On close examination, however—and whatever may be the other faults of the study (some of which are quite significant)—we find that it does not escape the trap of inadvertently treating the "system" as if it were just another "environment" to be controlled. In other words, implicit in the study is the unfounded assumption that the "controls" are somehow external to the "system" (e.g., the "political-legislative" process is somehow separate from the basis economic process, and not subordinate to it). In this sense, *The Limits to Growth* is a striking example of what we may now label the "knob-twiddling" approach to the political economy of socioeconomic change. Without meaning to, it unconsciously perpetuates what are in essence *ideological* illusions about the locus of control in complex adaptive systems. Correlatively, it reinforces equivalent illusions about the real role of ideas, policies, and "planning" in human systems. This type of approach effectively limits itself to a search for the socioeconomic equivalents of the "thermostats" in the system (the black boxes which are assumed to control it). The implication is that once these "controllers" are found—or, in other versions of the same approach, once the ones society believes in have been properly influenced by scientific reasoning—then their sensitivity, their design(s), and/or their "value-settings" can be modified so that the output of the system will be more appropriately distributed, and the crisis will pass.

What is left out of the analysis of the system in such studies, and what is missing from the various policies they put forward, is the starkly simple fact that the envisaged or actual "controllers" have a merely *functional* relationship to the system as a whole. As in any known society, these functional controls are subordinate to already-given *structural constraints*. The "controllers" operate at the level of the *messages* which constitute the surface structure of the system, whereas the structural constraints operate at the level of its deep structure, the level of its constraining *codes*. We do not know, in any long-term sense, just what

---

with the "survival of the fittest" and the "dream of the chance of 'success'"—in our society. But as a statement about economics, the analogy with Monopoly is worse than misleading. Precisely because we consider it "only a game," we usually fail to see that Monopoly supports the ideology of competition by basing itself on a logical and ecological absurdity. It is assumed that the winning player, having consumed all the resources of all the opponents, can actually survive the end of the game. In fact this is impossible. Unlike a true predator, which never competes with its prey, the Monopoly winner is "unfit" and must consequently die because in the context of the resources provided by the game, the winner has consumed them all, leaving no environment at all (no other players) to feed on.

[42]Donella Hager Meadows et al., *The Limits to Growth: A Report for the Club of Rome's Project on the Predicament of Mankind* (New York: Universe Books, 1972).

contribution functional adjustments in the surface structure of societies do in fact make to deep-structure (morphological) change. Little current historical knowledge is of help here, partly because the two levels of change go hand in hand (if not in synchrony), and partly because most of the history we were socialized in shares the same surface-structure (and "morphostatic") approach.

All that can be said with any certainty at this point is that functional adjustments belong to the domain of the symptoms of crises, not to the domain of the real problems which generate them. And this, in the last analysis, is the crux of the entire issue, for—if we may borrow an image from epidemiology—it is a standard axiom in the treatment of illness that the only time it is legitimate to treat symptoms, rather than the disease, is when the disease is beyond a cure.

The Power of Technique and The Ethics of Non-Power

JACQUES ELLUL

The problem of ethics and technique may be stated as follows: "in its concrete applications, technology raises a certain number of moral problems to which a solution must be sought." Euthanasia, non-human language, artificial life-support systems, psychological and genetic experimentation and research are cases in point. This is the traditional way of posing the problem, but it is no longer satisfactory today, for it serves to maintain a certain double status quo, by suggesting:

1) that our world has not changed, it has simply acquired technology, which must be treated as a separate issue; and
2) that the moral code has not changed either.

Ethics is thus split into a general system, on the one hand, and its application in specific instances on the other: euthanasia and abortion, for example, to which ethical principles that are deemed permanent, the product of a stable society, are applied. I believe, on the contrary, that a profound upheaval has taken place.

Instead of being merely a concrete element incorporated in a certain number of objects, technique can be abstract, and furthermore, instead of being a secondary factor which has been integrated into a stable civilization, technique has become the determining factor in all the problems which we face. Technique has likewise become a generalized mediation, so that it is no longer possible for us to form relationships of any kind without its intervening between us and our environment. Indeed technique itself has become an environment, replacing the natural one—witness the increasing numbers of people who live in cities where everything is either the product of technology or part of a technological process.

Technique proceeds in a causal, never in a goal-oriented fashion. It advances as a result of pre-existing techniques, which combine to facilitate a step forward, that is all. And this technological nexus is characteristically ambivalent. Because the solution of one problem by technological means immediately raises a multitude of others, which result directly from those very means, it is impossible to say whether technique produces good or evil effects. It does both, simultaneously. We are confronted with a system in the strict sense of the word, what I will call the "technological system," hence ethical issues may only be considered relative to the system as a whole, and not to specific instances. Because of the systemic nature of technique, it cannot be neutral. Hence, to claim that, for example, "technique is simply a knife. You can use it to cut up bread, or your neighbor, it is simply a question of use" is quite mistaken. The difference in power between a space-rocket and a knife makes a qualitative difference between the two inevitable.

Once we realize that technique is not a mere instrument of our will, a tool which we can use according to whim, our conviction that man remains in control is undermined. As soon as one asks "who controls?" it becomes apparent that although I may control my tape-recorder, or my television, for example, by not using them, not even the technician himself (who is inevitably a specialist) controls the entire technological system. As to the nature of the man who is believed to be in control, he is not, contrary to popular belief, the same as in the age of Pericles: he has already been molded by technology. As a result, the time-honored moral positions have become completely outdated. It no longer makes sense to attempt to distinguish between personal and social ethics. For a long time, the proposed solution to this dilemma was the famous theory of *adiaphora*, in other words (for the benefit of non-technicians in ethics), questions which concerned neither good nor evil, neutral questions: the delight of theologians. There were, so the argument went, problems of good and evil, and in between, issues which were neither good nor bad, hence there *was* no ethical problem. The reality, however, is the insidious ethics of adaptation, which rests on the notion that since technique is a fact, we should adapt ourselves to it. Consequently, anything that hinders technique ought to be eliminated, and thus adaptation itself becomes a moral criterion.

The development of technique has thus resulted in a new morality, technological morality, which has two characteristics:

1) it is behavioral (in other words, only correct practice, not intentions or motivations, counts), and

2) it rules out the problematics of traditional morality (the morality of ambiguity is unacceptable in the technological world).

Technique itself has become a virtue and (paralleling the scientific community's attempt to found a morality on scientific integrity) proposes the values of normalcy, efficiency, industriousness, professional ethics, and devotion to collective projects as values. In each case, everything is subordinated to efficiency, in other words, geared towards adaptation. Hence technological morality consists in allowing technique free play, and if traditional values are invoked, it is usually for another reason than to justify the primacy of technique (B.F. Skinner's well-known work is entirely representative of the morality of the technological era, which rules out not only traditional values, but certain modes of behavior as well. Thus, within technological morality, laziness is clearly unacceptable, waste is scandalous, and playing is merely for children). If, however, one seeks a common denominator for the value system proposed by technique and the behavior which it demands, it becomes clear that the real issue is power. All technique is a function of power, and even if we focus on specific cases or hypotheses, we realize that it is always because man has the power to do almost anything that fragmentary problems arise. That power, however, is not man's, it remains extrinsic to him. It is exclusively concerned with means and it is the excessiveness of these means which is ultimately the cause of the crisis in our civilization and in our system of ethics. Whereas the latter was originally formulated for men without technical resources (hence every problem was one of *direct* control and intention), now it is a question of resources and power.

At the level of power, the first essential factor is the established fact that there is a contradiction between power and values. Every increase in power ends in a challenge to, or a defeat of values (this is a pragmatic proposition). But if values are called into question, no conceivable limits, no benchmarks by which conduct may be evaluated, remain: man becomes incapable of exercising judgment, since his judgment depends on values. The only remaining rule is that "everything that can be done, ought to be done." Power always implies a plus, an "in addition to"; in order to pose, accept, and respect limits, some commonly accepted values are necessary.

But the problem of power is not simply the result of a certain will to power. Power is not autonomous. It exists today only as a result of means, it is inscribed in a world of means. Ends and means can no longer be separated—they are interdependent, defined by each other—but it is always technique which supplies the means, whose power and thrust dominate the entire field of contemporary thought and life. Thus, if we want to assess accurately the problems of ethics today and guide the direction of research in ethics, it must be in the context of this growth of power and this universe of means. Here we must take our stand and not, as many are currently prone to do, in a universe of hypothetical ends. (The passion for utopias represents precisely the evasion of our current problems; we look very far ahead, contemplating the year 2050 when all of them have been solved. I, however, am concerned with the period between 1980 and 2000; this is the important moment, and it is not a utopian one.)

In this technological society, we must also seek an ethics, which would play the traditional role of ethics, that of preserving man's control over life, providing, for example, the possiblity of the development of society and personal relationships. Hence, if we continue to accept this dual orientation of ethics, one crucial direction which research in contemporary ethics should take would concern technique, without, however, being anti-technique. For we cannot claim to be anti-technique, we are deeply implicated in it and cannot be otherwise.

The ethics I have in mind would have four characteristics: it would be an ethics of non-power, freedom, conflict, and transgression. This is not an original idea of mine; most current research on technique points in this direction, as, for example, when Bernard de Jouvenel speaks of amenity, Ivan Illich of conviviality, Georges Friedmann of wisdom, Jean Fourastié of necessity and personal discipline, not to mention Denis de Rougement and Jean-Pierre Domenach. It is, in every case, a question of some form of reduction of power: that man accept *not* to do all that he is capable of doing. The logic of technique, on the other hand, demands that whatever can be done *must* be done. Yet when I speak of the ethics of non-power, I do not mean impotence. Non-power does not mean giving something up, but choosing *not* to do something, being capable of doing something and deciding against it. Nor, by the same token, is it fatalism. It is an ethic which operates at every level, including the level of personal behavior in everyday life (adopting an attitude of non-power, for example, when one is driving a car or when one has a transistor radio which is too loud for the neighbors). It involves a permanent decision which is not only personal but also institutional, because it challenges manipulation and automatic growth; it is both a refusal of competition and the institution of a new, non-competitive pedagogy. The ethics of non-power has the effect of calling into question such events as the Olympic Games, automobile racing, and so on, but it is also highly relevant to scientific research. What is at stake is a vital principle (I simplify here to make a point) of setting limits: given that the almost unlimited means at our disposal permit almost unlimited action, *we must choose, a priori, non-intervention each time there is uncertainty about the global and long-term effects of whatever actions are to be undertaken.* This ethics, this opting for non-power is fundamental, and it is possible (it would be futile to formulate an ideal ethics which no one could practice) because it is linked with meaning. Our experience with the power of technique has led us to discover the absence of meaning. Uncertainty as to whether life means anything is the sickness of modern man, and the rediscovery of meaning is conditional upon the choice of non-power.

This new ethics would also be an ethics of freedom. Powerful means do not necessarily insure freedom; on the contrary, technique has come to represent both necessity and fate for modern man, and thus, the effort to recover our ethical identity is the equivalent of resuming the fight for freedom. Not that I believe man is free; I insist, on the contrary, that man is determined, and has always been, but that man claims to be free, wishes to be free, affirms his freedom, and fights for it. This process has had three stages. Initially man was heavily determined by the forces of nature, from whose bonds he struggled successfully to free himself, winning, with the aid of technique, a high degree of independence. But at the point where man began to conquer nature, he found himself heavily determined by society, and his response to social determination was revolution. Now it is technique which determines man, but surely the technological system is no stronger or more dangerous than nature was for prehistoric man. Whereas prehistoric man discovered that useful tool—technique—for the development of a value system and a symbolic order, ethics is also a useful instrument for achieving liberation, but it must be carefully chosen. In other words, we must decide (and the decision carries with it grave consequences) that it is not technique which frees us but rather it is from technique that we must free ourselves. Our experience in this context is similar to that of the youth of every generation, who long for freedom, but do not quite know who their adversaries are. To fight against a well-defined enemy, such as Hitler, is comparatively easy. Now we are threatened by obscure and diffuse powers and are fighting in the dark against unfamiliar forces which we have not been able to analyze. Adopting an analytical perspective on the technological system should help to dispel that darkness.

Since the principles of non-power and freedom necessarily create conflict and tension, the ethics I propose would also include the principle of conflict. It is a matter of re-introducing conflict and play, of making holes in a social fabric which technique would wish to remain seamless. Technique is unifying and totalizing, whereas a group of people can exist only in conflict and negotiation. The moment a perfectly homogeneous group of people is achieved, we no longer exist, either collectively or individually. Thus conflict is a fundamental ethical value (clearly we have come a long way from traditional morality).

Finally, this ethics should be one of transgression and profanation. But here we must be prudent, because there must be no mistake about what should be transgressed. Today everybody transgresses sexual taboos, drug laws, and so on, but these transgressions are meaningless and do not in any way constitute a challenge to the constraints of society. Such a challenge can only be posed by transgressing the constraints imposed by technique, in other words, what is real. This can only be achieved by demythologizing and desacralizing technique, in spite of the blind faith we all place in it. Especially, I believe, we must destroy the illusion of progress, the illusion that technique leads us from one achievement to another, the deep-rooted illusion that the material and the spiritual coincide. We tend to think that technique liberates us from the mundane, from material needs, so that we become free-floating pure spirits. But alas, technique, while it liberates us from one thing deprives us of something else at the same time, and that something else is usually of the spiritual order.

Every artist knows that he must overcome resistance; if technique overcomes the resistance of his materials for him, he can no longer conceive of anything. In order to create, I need to meet a resistance which technique ought not to deprive me of. I am neither liberated nor dematerialized by technique. In other words, technique ought to be reduced to producing merely useful objects, which *function*. When a new technique becomes available to me, although I may not understand it, I like it to work. It is useful, no more. But does such usefulness warrant the sacrifices which are demanded from us? That is finally the question we must face. It is this transgression of the technological ideology that we have absorbed, which allows for the establishment of new limits, such as those sought by Illich.

In conclusion, I should like to say that what I propose is neither trivial, retrograde nor destructive of technique, it is simply an attempt to deal with a new environment which we do not know very well. It is a matter of reaffirming ourselves as subjects, and I believe that insofar as we speak, we are still subjects. Neither this reaffirmation nor the raising of ethics as an issue is opposed either to man or society, but is directed towards keeping both alive. That is the task of ethics: a task to which, understandably, time-honored values are no longer quite equal. Such values are, however, irreplaceable, because there are no substitutes for freedom and dignity, and there I will rest my case.

Translated by Mary Lydon
University of Wisconsin-Milwaukee

Contributors

JEAN BAUDRILLARD is Professor of Sociology at the University of Paris-Nanterre. He pursues his analysis of contemporary social life in the periodical *Traverse*, as well as in his books, *Système des objets* (1968), *Pour une critique de l'économie politique du signe* (1972), and *L'Echange symbolique et la mort* (1976).

JACK BURNHAM is Professor of Art at Northwestern University. He is the author of *Beyond Modern Sculpture: The Effects of Science and Technology on the Sculpture of This Century* (1968) and of numerous articles, among which "Systems Esthetics," *Artforum* (May 1969), and "Systems and Art," *Arts in Society* (Summer/Fall 1969) are particularly relevant to the present context.

DANIEL CHARLES is Director of the Department of Music at the University of Paris-Vincennes. He has written extensively on contemporary experimental music, notably on the work of John Cage, and his recent publications include *Gloses sur John Cage* (1978) and *Le Temps de la voix* (1978).

MIKEL DUFRENNE teaches at the University of Paris-Nanterre and has written extensively on aesthetics and the relationship between art and society. His works include *Art et Politique* (1974), *Esthétique et philosophie* 1 (1967) and 2 (1976), and *Phénoménologie de l'experience esthétique*, which appeared in English as *The Phenomenology of Aesthetic Experience* (1973).

248

JEAN-PIERRE DUPUY is Professor of Economics at L'Ecole Polytechnique in Paris. He is the author of *L'Invasion pharmaceutique* (1974), *Valeur sociale et l'encombrement du temps* (1975) and *La Trahison de l'opulence* (1976).

JACQUES ELLUL is Professor of the History of Institutions at the University of Bordeaux 1. He is the author of numerous works on various aspects of contemporary sociology and theology, among the most recent of which are: *L'Espérance oubliée* (1977) and *Trahison de l'occident* (1975). *La Technique; ou l'enjeu du siècle* has appeared in English under the title *The Technological Society* (1964).

ANDREW FEENBERG teaches philosophy at San Diego State University. Among his recent publications is a chapter entitled "Transition or Convergence: Communism and the Paradox of Development," in *Technology and Communist Culture*, ed. Frederick J. Fleron Jr. (1977). He has co-authored with James Freedman a book on the French May 1968 events, *The Changing Face of Socialism*, forthcoming from South End Press.

HEINZ VON FOERSTER is Professor Emeritus of the Departments of Biophysics and of Electrical Engineering of the University of Illinois, where he established the Biological Computer Laboratory in 1959. His many publications include "Thoughts and Notes on Cognition," in *Cognition, a Multiple View*, ed. P.L. Garvin (1970), and "Notes pour une épistemologie des objets vivants" in *L'Unité de l'homme*, eds. E. Morin and M. Piattelli-Pamarini (1974).

DAVID HALL is Professor of Philosophy at the University of Texas at El Paso. His book, *The Civilization of Experience* (1973), develops a theory of cultural interpretation which he has applied to contemporary technological culture in his forthcoming *The Uncertain Phoenix: Adventures Towards a Post-Cultural Sensibility*.

ANDREAS HUYSSEN is Associate Professor of German and Comparative Literature at the University of Wisconsin-Milwaukee. He is the author of *Die fruhromantische Konzeption von Übersetzung und Aneignung: Studien zur fruhromantischen Utopie einer deutschen Weltliteratur* (1969), *Bürgerlicher Realismus* (1974) and *Das Drama des Sturm und Drang* (1979) and articles on twentieth-century art, literature, and cultural theory, some of which have appeared in *New German Critique*, of which he is an editor.

ERIC LEED is Associate Professor of History at Florida International University. He was a Fellow of the National Humanities Institute on Technology and the Humanities in 1976–77. His current research focuses on the history of travel, and his publications include *No Man's Land: Combat of Identity in World War I* (1979) and "Communications Revolution and the Enactment of Culture," *Communications Research*, 5 (1978).

MAGOROH MARUYAMA is Visiting Professor in the Department of Administrative Sciences at Southern Illinois University at Carbondale. His numerous publications include *Culture of the Future* (1978) and "Heterogenistics and Morphogenistics," in *Theory and Society*, 5 (1978).

OSKAR NEGT is Professor of Social Studies at the Technische Universität, Hannover. He is the author of numerous articles, many of which have been published in English translation in *Telos* and *New German Critique*, and of several books, among which are *Soziologische Phantasie und exemplarisches Lernen* (1968), *Politik als Protest* (1971), and *Öffentlichkeit und Erfahrung* (1972).

ANTHONY WILDEN is Professor of Communications at Simon Fraser University. His publications include *The Language of the Self*, with Jacques Lacan (1968), *System and Structure: Essays in Communication and Exchange* (1972; 2nd. ed., 1980), and *The Imaginary Canadian* (1980).

KATHLEEN WOODWARD is Assistant Professor of English and Cultural and Technological Studies at the University of Wisconsin-Milwaukee. Her publications include *At Last the Real Distinguished Thing: The Late Poems of Eliot, Pound, Stevens, and Williams* (forthcoming 1980), and, as co-editor, *Aging and the Elderly: Humanistic Perspectives in Gerontology* (1978).

JACK ZIPES is Professor of German and Comparative Literature at the University of Wisconsin-Milwaukee. He is co-editor of *New German Critique* and contributing editor of *Theater*. His books include *The Great Refusal: Studies of the Romantic Hero in German and American Literature* (1971), *Political Plays for Children* (1976), and *Breaking the Magic Spell: Radical Theories of Folk and Fairy Tales* (1979).